THE COST OF LIVING

ARUNDHATI ROY

THE COST OF LIVING

The Greater Common Good

AND

The End of Imagination

Flamingo
An Imprint of HarperCollins*Publishers*

Flamingo
An Imprint of HarperCollins*Publishers*
77–85 Fulham Palace Road,
Hammersmith, London w6 8jb

www.**fire**and**water**.com

Published by Flamingo 1999

Flamingo is a registered trademark
of HarperCollins *Publishers* Limited

1 3 5 7 9 8 6 4 2

The Greater Common Good was first published in book form
in India in 1999 by IBD. An earlier version of The End of
Imagination was first published in *Outlook* and *Frontline*
magazines in India 1998

A catalogue record for this book
is available from the British Library

isbn 0 00 257187 0

Typeset in Postscript Janson by
Rowland Phototypesetting Ltd,
Bury St Edmunds, Suffolk

Printed and bound in Great Britain by
Clays Ltd, St Ives plc

PREFACE

In May 1998, the government of India conducted
a series of nuclear tests in the Thar desert and
declared itself a nuclear weapons state. Within days
the Government of Pakistan followed suit.

I was in the United States on a reading tour
of my book, *The God of Small Things*, when this
happened. My first response was one of disgust at
the condescension, the hypocrisy and the double
standards of the reaction in the western world. (Can
the Blacks handle the Bomb?)

I returned to India and it took a few months for
me to stop reacting to the international reaction
and to begin to address what we had done to ourselves, to our lives, to our futures. In July 1998, I
wrote 'The End of Imagination'. In August, it was
published simultaneously in two mainstream magazines – *Frontline* and *Outlook* – as a cover story.
Both publications set aside competing commercial
considerations and did this despite the cacophony
of nationalism and jingoism that was (and still is)
being orchestrated by political parties and much of
the press.

Nuclear bombs, we were told, were necessary as
a deterrent. A deterrent to what?

Today, a year after the nuclear tests, the hostility between India and Pakistan has spiralled into a flashpoint. We're not supposed to call it war. But both countries are counting their dead.

'Oh, but it's only a conventional war,' I was told by a senior journalist who was interviewing me.

Only a conventional war?

Have we raised the threshold of horror so high that nothing short of a nuclear strike qualifies as a *real* war? Are we to spend the rest of our lives in this state of high alert with guns pointed at each other's heads and fingers trembling on the trigger?

Thank you, Government of India, thank you, Government of Pakistan. But most of all, thank you, Government of the U.S. of A.

We're deeply, deeply grateful.

The second essay, 'The Greater Common Good' (the first in this book), was also published simultaneously in *Outlook* and *Frontline* in June 1999.

In February 1999, the newspapers reported that the Supreme Court of India had lifted a four-year-long legal stay on the construction of a controversial megadam – the Sardar Sarovar – being built on the Narmada river in central India. It is one of 3,200 dams being planned on a single river. Its proponents boast that this is the largest, most ambitious, river valley development project ever conceived in human history.

I began to follow the story. The more I read, the more horrified I became. In March I travelled to the Narmada valley. I returned, numbed. I returned unable to ignore or accept what everybody (including myself) has, over the years, gradually accepted and successfully ignored.

India is the third largest dam builder in the world. In the last fifty years (since Independence), India has built 3,300 Big Dams. Their reservoirs have uprooted millions of people. *Yet there are no government records of how many people have been displaced.* India does not even have a national rehabilitation policy.

These thousands of dams have been built in the name of National Development. Yet 250 million people have no access to safe drinking water. At least 350 million people (more than the country's population at the time of Independence) live below the poverty line. Over eighty per cent of rural households do not have electricity.

Geographically, there has been an *increase* in flood-prone and drought-prone areas since 1947!

The government – every Indian government – refuses to address the problem. To even *consider* that something is amiss.

Even as I write, the monsoon is raging outside my window. It's high noon, but the sky is dark, and my lights are on. I know that the waters of the Sardar Sarovar reservoir are rising every hour. More than ten thousand people face submergence. They have nowhere to go. I have tried very hard to communicate the urgency of what is happening in the valley. But in the cities, peoples' eyes glaze over. 'Yes, it's sad,' we say. 'But it can't be helped. We need electricity.' The story of the Narmada valley is nothing less than the story of Modern India. Like the tiger in the Belgrade Zoo during the NATO bombing, we've begun to eat our own limbs.

Arundhati Roy
July 1999

The Greater Common Good

ACKNOWLEDGEMENTS

There are two men who fall into the *without whom* (this essay couldn't have been written) category:

Himanshu Thakker, who first revealed to me – brilliantly, meticulously, almost shyly – the horrors of the Narmada Valley Development Projects. To him I owe my first (belated) conspectus of this intricate method of pulverising a people.

Patrick McCully, who I've never met, but whose book *Silenced Rivers* is the rock on which this work stands. If you want to read a truly dazzling book on Big Dams, drop mine and read his.

Jharana Jhaveri, most tenacious of fighters and gentlest of friends, thank you for travelling with me. All the way.

Shripad, Nandini, Silvie, Alok, Medha, Baba Amte and their colleagues at NBA. Extraordinary people, fighting an extraordinary war.

Deepak Sarkar and Anurag Singh, for your friendship and cool, organized wisdom and advice.

N. Ram and Vinod Mehta, editors of *Outlook* and *Frontline*, who first published *The Greater Common Good*. There aren't many like you around.

Jojo Van Gruisen, Golak Khandual, Arjun Raina, Sanjay Kak. Old Soul friends. Fellow travellers on this path.

Finally, Pradip Krishen, without whom my life would not be fully lived.

Thank you.

To
the Narmada,
and all the life she sustains

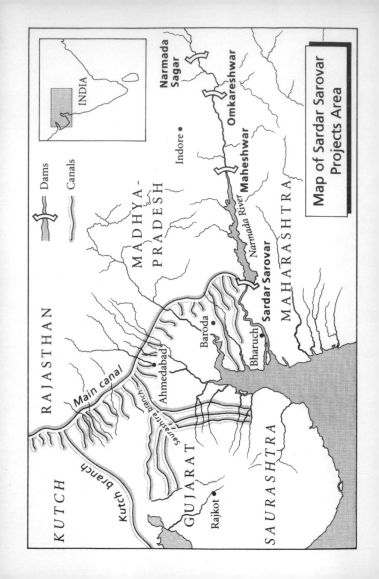

Map of Sardar Sarovar Projects Area

'If you are to suffer, you should suffer in the interest of the country . . .'

Jawaharlal Nehru, speaking to villagers who were to be displaced by the Hirakud dam, 1948[1]

I stood on a hill and laughed out loud.

I had crossed the Narmada by boat from Jalsindhi and climbed the headland on the opposite bank from where I could see, ranged across the crowns of low, bald hills, the Adivasi* hamlets of Sikka, Surung, Neemgavan and Domkhedi. I could see their airy, fragile homes. I could see their fields and the forests behind them. I could see little children with littler goats scuttling across the landscape like motorized peanuts. I knew I was looking at a civilization older than Hinduism, slated – *sanctioned* (by the highest court in the land) – to be drowned this monsoon when the waters of the Sardar Sarovar reservoir will rise to submerge it.

* 'Adivasi' is the term now used in India to designate the original inhabitants (indigenous people) of a region.

Why did I laugh?

Because I suddenly remembered the tender concern with which the Supreme Court judges in Delhi (before vacating the legal stay on further construction of the Sardar Sarovar dam) had enquired whether Adivasi children in the resettlement colonies would have children's parks to play in. The lawyers representing the Government had hastened to assure them that indeed they would, and what's more, that there were seesaws and slides and swings in every park. I looked up at the endless sky and down at the river rushing past and for a brief, brief moment the absurdity of it all reversed my rage and I laughed. I meant no disrespect.

Let me say at the outset that I'm not a city-basher. I've done my time in a village. I've had first-hand experience of the isolation, the inequity and the potential savagery of it. I'm not an anti-development junkie, nor a proselytiser for the eternal upholding of custom and tradition. What I *am*, however, is curious. Curiosity took me to the Narmada valley. Instinct told me that this was the big one. The one in which the battle-lines were clearly drawn, the warring armies massed along them. The one in which it would be possible to wade through the congealed morass of hope, anger, information, disinformation, political artifice, engineering ambition, disingenuous socialism, radical activism, bureaucratic subterfuge, misinformed emotionalism and of course the pervasive, invariably dubious, politics of International Aid.

Instinct led me to set aside Joyce and Nabokov, to postpone reading Don DeLillo's big book and substitute it with reports on drainage and irrigation, with journals and books and documentary films about dams and why they're built and what they do. My first tentative questions revealed that few people know what is really going on in the Narmada valley. Those who know, know a lot. Most know nothing at all. And yet, almost everyone has a passionate opinion. Nobody's neutral. I realized very quickly that I was straying into mined territory.

In India over the last ten years the fight against the Sardar Sarovar dam has come to represent far more than the fight for one river. This has been its strength as well its weakness. Some years ago, it became a debate that captured the popular imagination. That's what raised the stakes and changed the complexion of the battle. From being a fight over the fate of a river valley it began to raise doubts about an entire political system. What is at issue now is the very nature of our democracy. Who owns this land? Who owns its rivers? Its forests? Its fish? These are huge questions. They are being taken hugely seriously by the State. They are being answered in one voice by every institution at its command – the army, the police, the bureaucracy, the courts. And not just answered, but answered unambiguously, in bitter, brutal ways.

For the people of the valley, the fact that the stakes were raised to this degree has meant that their most effective weapon – *specific* facts about *specific* issues in this *specific* valley – has been blunted by the debate on the big issues. The basic premise of the argument has been inflated until it has burst into bits that have, over time, bobbed away. Occasionally a disconnected piece of the puzzle floats by – an emotionally charged account of the Government's callous treatment of displaced people; an outburst at how the Narmada Bachao Andolan (NBA), 'a handful of activists', is holding the nation to ransom; a legal correspondent reporting on the progress of the NBA's writ petition in the Supreme Court.

Though there has been a fair amount of writing on the subject, most of it is for a 'special interest' readership. News reports tend to be about isolated aspects of the project. Government documents are classified as 'Secret'. I think it's fair to say that public perception of the issue is pretty crude and is divided crudely, into two categories:

On the one hand, it is seen as a war between modern, rational, progressive forces of 'Development' vs a sort of neo-Luddite impulse – an irrational, emotional 'Anti-Development' resistance, fuelled by an arcadian, pre-industrial dream.

On the other, as a Nehru vs Gandhi contest. This lifts the whole sorry business out of the bog of deceit, lies, false promises and increasingly successful propaganda (which is what it's *really* about) and confers on it a false legitimacy. It makes out that both sides have the Greater Good of the Nation in mind – but merely disagree about the means by which to achieve it.

Both interpretations put a tired spin on the dispute. Both stir up emotions that cloud the particular facts of this particular story. Both are indications of how urgently we need new heroes – new *kinds* of heroes – and how we've overused our old ones (like we overbowl our bowlers).

The Nehru vs Gandhi argument pushes this very contemporary issue back into an old bottle. Nehru and Gandhi were generous men. Their paradigms for development are based on assumptions of inherent morality. Nehru's on the paternal, protective morality of the Soviet-style centralized State. Gandhi's on the nurturing, maternal morality of romanticized village Republics. Both would probably work, if only we were better human beings. If we all wore khadi and suppressed our base urges. Fifty years down the line, it's safe to say that we haven't made the grade. We haven't even come close. We need an updated insurance plan against our own basic natures.

It's possible that as a nation we've exhausted our quota of heroes for this century, but while we wait for shiny new ones to come along, we have to limit the damage. We have to support our small heroes. (Of these we have many. Many.) We have to fight specific wars in specific ways. Who knows, perhaps that's what the twenty-first century has in store for us. The dismantling of the Big. Big bombs, big dams, big ideologies, big contradictions, big countries, big wars, big heroes, big mistakes. Perhaps it will be the Century of the Small. Perhaps right now, this very minute, there's a small god up in heaven readying herself for us. Could it be? Could it *possibly* be? It sounds finger-licking good to me.

I was drawn to the valley because I sensed that the fight for the Narmada had entered a newer, sadder phase. I went because writers are drawn to stories the way vultures are drawn to kills. My motive was not compassion. It was sheer greed. I was right. I found a story there.

And what a story it is . . .

'People say that the Sardar Sarovar dam is an expensive project. But it is bringing drinking water to millions. This is our lifeline. Can you put a price on this? Does the air we breathe have a price? We will live. We will drink. We will bring glory to the state of Gujarat.'

Urmilaben Patel, wife of Gujarat Chief Minister
Chimanbhai Patel, speaking at a public rally in
Delhi in 1993.

'We will request you to move from your houses after the dam comes up. If you move it will be good. Otherwise we shall release the waters and drown you all.'

Morarji Desai, speaking at a public meeting in
the submergence zone of the Pong dam in 1961.[2]

'Why didn't they just poison us? Then we wouldn't have to live in this shit-hole and the Government could have survived alone with its precious dam all to itself.'

Ram Bai, whose village was submerged when
the Bargi dam was built on the Narmada. She
now lives in a slum in Jabalpur.[3]

In the fifty years since Independence, after Nehru's famous 'Dams are the Temples of Modern India' speech (one that he grew to regret in his own life-time[4]), his footsoldiers threw themselves into the business of building dams with unnatural fervour. Dam-building grew to be equated with Nation-building. Their enthusiasm alone should have been reason enough to make one suspicious. Not only did they build new dams and new irrigation systems, they took control of small, traditional systems that had been managed by village communities for thousands of years, and allowed them to atrophy.[5] To compensate the loss, the Government built more and more dams. Big ones, little ones, tall ones, short ones. The result of its exertions is that India now boasts of being the world's third largest dam builder. According to the Central Water Commission, we have 3,600 dams that qualify as Big Dams, 3,300 of them built after Independence. One thousand more are under construction.[6] Yet one-fifth of our population – 200 million people – does not have safe drinking water and two-thirds – 600 million – lack basic sanitation.[7]

Big Dams started well, but have ended badly. There was a time when everybody loved them, everybody had them – the Communists, Capitalists, Christians, Muslims, Hindus, Buddhists. There was a time when Big Dams moved men to poetry. Not any longer. All over the world there is a movement growing against Big Dams.

In the First World they're being de-commissioned, blown up.[8] The fact that they do more harm than good is no longer just conjecture. Big Dams are obsolete. They're uncool. They're undemocratic. They're a Government's way of accumulating authority (deciding who will get how much water and who will grow what where). They're a guaranteed way of taking a farmer's wisdom away from him. They're a brazen means of taking water, land and irrigation away from the poor and gifting it to the rich. Their reservoirs displace huge populations of people, leaving them homeless and destitute.

Ecologically too, they're in the doghouse.[9] They lay the earth to waste. They cause floods, waterlogging, salinity, they spread disease. There is mounting evidence that links Big Dams to earthquakes.

Big Dams haven't really lived up to their role as the monuments of Modern Civilisation, emblems of Man's ascendancy over Nature. Monuments are supposed to be timeless, but dams have an all too finite lifetime. They last only as long as it takes Nature to fill them with silt.[10] It's common knowledge now that Big Dams do the opposite of what their Publicity People say they do – the Local Pain for National Gain myth has been blown wide open.

For all these reasons, the dam-building industry in the First World is in trouble and out of work. So it's exported to the Third World in the name of Development Aid,[11] along with their other waste like old weapons, superannuated aircraft carriers and banned pesticides.

On the one hand the Indian Government, *every* Indian Government, rails self-righteously against the First World, and on the other, actually *pays* to receive their gift-wrapped garbage. Aid is just another praetorian business enterprise. Like Colonialism was. It has destroyed most of Africa. Bangladesh is reeling from its ministrations. We *know* all this, in numbing detail. Yet in India our leaders welcome it with slavish smiles (and make nuclear bombs to shore up their flagging self-esteem).

Over the last fifty years India has spent Rs 87,000 crores[12] on the irrigation sector alone.[13] Yet there are more drought-prone areas and more flood-prone areas today than there were in 1947.[14] Despite the disturbing evidence of irrigation disasters, dam-induced floods and rapid disenchantment with the Green Revolution[15] (declining yields, degraded land), the Government has not commissioned a post-project evaluation of a *single one* of its 3,600 dams to gauge whether or not it has achieved what it set out to achieve, whether or not the (always phenomenal) costs were justified, or even what the costs actually were.

The Government of India has detailed figures for how many million tonnes of food grain or edible oils the country produces and how much more we produce now than we did in 1947. It can tell you how much bauxite is mined in a year or what the total surface area of the National Highways adds up to. It's possible to access minute-to-minute information about the stock exchange or the value of the rupee in the world market. We know how many cricket matches we've lost on a Friday in Sharjah. It's not hard to find out how many graduates India produces, or how many men had vasectomies in any given year. But the Government of India does not have a figure for the number of people that have been displaced by dams or sacrificed in other ways at the altars of 'National Progress'. Isn't this *astounding*? How can you measure Progress if you don't know what it costs and who has paid for it? How can the 'market' put a price on things – food, clothes, electricity, running water – when it doesn't take into account the *real* cost of production?

According to a detailed study of fifty-four Large Dams done by the Indian Institute of Public Administration,[16] the *average* number of people displaced by a Large Dam in India is 44,182. Admittedly, fifty-four dams out of 3,300 is not a big enough sample. But since it's all we have, let's try and do some rough arithmetic. A first draft.

To err on the side of caution, let's halve the number of people. Or, let's err on the side of *abundant* caution and take an average of just 10,000 people per Large Dam. It's an improbably low figure, I know, but . . . never mind. Whip out your calculators. 3,300 × 10,000 = 33,000,000

That's what it works out to, thirty-three *million* people. Displaced by Big Dams *alone* in the last fifty years. What about those who have been displaced by the thousands of other Development Projects? At a private lecture, N. C. Saxena, Secretary to the Planning Commission, said he thought the number was in the region of 50 million (of whom 40 million were displaced by dams).[17] We daren't say so, because it isn't official. It isn't official because we daren't say so. You have to murmur it for fear of being accused of hyperbole. You have to whisper it to yourself, because it really does sound unbelievable. It *can't be*, I've been telling myself. I must have got the zeroes muddled. *It can't be true.* I barely have the courage to say it aloud. To run the risk of sounding like a 1960s hippie dropping acid ('It's the System, man!'), or a paranoid schizophrenic with a persecution complex. But it *is* the System, man. What else can it be?

Fifty million people.

Go on, Government, quibble. Bargain. Beat it down. Say *something*.

I feel like someone who's just stumbled on a mass grave.

Fifty million is more than the population of Gujarat. Almost three times the population of Australia. More than three times the number of refugees that Partition created in India. Ten times the number of Palestinian refugees. The Western world today is convulsed over the future of one million people who have fled from Kosovo.

A huge percentage of the displaced are Adivasis (57.6 per cent in the case of the Sardar Sarovar dam).[18] Include Dalits and the figure becomes obscene. According to the Commissioner for Scheduled Castes and Tribes it's about 60 per cent.[19] If you consider that Adivasis account for only 8 per cent and Dalits another 15 per cent of India's population, it opens up a whole other dimension to the story. The ethnic 'otherness' of their victims takes some of the pressure off the Nation Builders. It's like having an expense account. Someone else pays the bills. People from another country. Another world. India's poorest people are subsidizing the lifestyles of her richest.

Did I hear someone say something about the world's biggest democracy?

What has happened to all these millions of people? Where are they now? How do they earn a living? Nobody really knows. (Recently, the *Indian Express* had an account of how Adivasis displaced from the Nagarjunasagar dam Project are selling their babies to foreign adoption agencies.[20] The Government intervened and put the babies in two public hospitals where six infants died of neglect.) When it comes to Rehabilitation, the Government's priorities are clear. India does not *have* a National Rehabilitation Policy. According to the Land Acquisition Act of 1894 (amended in 1984) the Government is not legally bound to provide a displaced person with anything but a cash compensation. Imagine that. A cash compensation, to be paid by an Indian government official to an illiterate male Adivasi (the women get nothing) in a land where even the postman demands a tip for a delivery! Most Adivasis have no formal title to their land and therefore cannot claim compensation anyway. Most Adivasis – or let's say most small farmers – have as much use for money as a Supreme Court judge has for a bag of fertiliser.

The millions of displaced people don't exist any more. When history is written they won't be in it.

Not even as statistics. Some of them have subsequently been displaced three and four times – a dam, an artillery proof range, another dam, a uranium mine, a power project. Once they start rolling there's no resting place. The great majority is eventually absorbed into slums on the periphery of our great cities, where it coalesces into an immense pool of cheap construction labour (that builds more projects that displace more people). True, they're not being annihilated or taken to gas chambers, but I can warrant that the quality of their accommodation is worse than in any concentration camp of the Third Reich. They're not captive, but they redefine the meaning of liberty.

And still the nightmare doesn't end. They continue to be uprooted even from their hellish hovels by government bulldozers that fan out on clean-up missions whenever elections are comfortingly far away and the urban rich get twitchy about hygiene. In cities like Delhi, they run the risk of being shot by the police for shitting in public places – like three slum-dwellers were, not more than two years ago.

In the French Canadian wars of the 1770s, Lord Amherst exterminated most of Canada's Native Indians by offering them blankets infested with the smallpox virus. Two centuries on, we of the Real India have found less obvious ways of achieving similar ends.

The millions of displaced people in India are nothing but refugees of an unacknowledged war. And we, like the citizens of White America and French Canada and Hitler's Germany, are condoning it by looking away. Why? Because we're told that it's being done for the sake of the Greater Common Good. That it's being done in the name of Progress, in the name of the National Interest (which, of course, is paramount). Therefore gladly, unquestioningly, almost gratefully, we believe what we're told. We believe what it benefits us to believe.

Allow me to shake your faith. Put your hand in mine and let me lead you through the maze. Do this, because it's important that you understand. If you find reason to disagree, by all means take the other side. But please don't ignore it, don't look away. It isn't an easy tale to tell. It's full of numbers and explanations. Numbers used to make my eyes glaze over. Not any more. Not since I began to follow the direction in which they point.

Trust me. There's a story here.

*

It's true that India has progressed. It's true that in 1947, when Colonialism formally ended, India was food deficient. In 1950 we produced 51 million tonnes of food grain. Today we produce close to 200 million tonnes.[21]

It's true that in 1995 the state granaries were over-flowing with 30 million tonnes of unsold grain. It's also true that at the same time, 40 per cent of India's population – more than 350 million people – were living below the poverty line.[22] That's more than the country's population in 1947.

Indians are too poor to buy the food their country produces. Indians are being forced to grow the kinds of food they can't afford to eat themselves. Look at what happened in Kalahandi district in western Orissa, best known for its starvation deaths. In the drought of 1996, people died of starvation (sixteen according to the State, over one hundred according to the press).[23] Yet that same year rice production in Kalahandi was higher than the

national average! Rice was exported from Kalahandi district to the Centre.

Certainly India has progressed but most of its
people haven't. Our leaders say that we must have
nuclear missiles to protect us from the threat of
China and Pakistan. But who will protect us from
ourselves?

What kind of country is this? Who owns it? Who
runs it? What's going on?
It's time to spill a few State Secrets. To puncture
the myth about the inefficient, bumbling, corrupt,
but ultimately genial, essentially democratic, Indian
State. Carelessness cannot account for 50 million
disappeared people. Nor can Karma. Let's not
delude ourselves. There is method here, precise,
relentless and 100 per cent man-made.

The Indian State is not a State that has failed. It is
a State that has succeeded impressively in what it
set out to do. It has been ruthlessly efficient in the
way it has appropriated India's resources – its land,
its water, its forests, its fish, its meat, its eggs, its

air – and redistributed it to a favoured few (in return, no doubt, for a few favours). It is superbly accomplished in the art of protecting its cadres of paid-up elite, consummate in its methods of pulverizing those who inconvenience its intentions. But its finest feat of all is the way it achieves all this and emerges smelling sweet. The way it manages to keep its secrets, to contain information – that vitally concerns the daily lives of one billion people – in government files, accessible only to the keepers of the flame: ministers, bureaucrats, state engineers, defence strategists. Of course we make it easy for them, we, its beneficiaries. We take care not to dig too deep. We don't really *want* to know the grisly detail.

Thanks to us, Independence came (and went), elections come and go, but there has been no shuffling of the deck. On the contrary, the old order has been consecrated, the rift fortified. We, the Rulers, won't pause to look up from our groaning table. We don't seem to know that the resources we're feasting on are finite and rapidly depleting. There's cash in the bank, but soon there'll be nothing left to buy with it. The food's running out in the kitchen. And the servants haven't eaten yet.

Actually, the servants stopped eating a long time ago.

India lives in her villages, we're told, in every other sanctimonious public speech. That's bullshit. It's just another fig leaf from the Government's bulging wardrobe. India doesn't live in her villages. India *dies* in her villages. India gets kicked around in her villages. India lives in her cities. India's villages live only to serve her cities. Her villagers are her citizens' vassals and for that reason must be controlled and kept alive, but only just.

This impression we have of an overstretched State, struggling to cope with the sheer weight and scale of its problems, is a dangerous one. The fact is that it's *creating* the problem. It's a giant poverty-producing machine, masterful in its methods of pitting the poor against the very poor, of flinging crumbs to the wretched so that they dissipate their energies fighting each other, while peace (and advertising) reigns in the Master's Lodgings.

Until this process is recognised for what it is, until it is addressed and attacked, elections – however fiercely they're contested – will continue to be mock battles that serve only to further entrench unspeakable inequity. Democracy (our version of it) will continue to be the benevolent mask behind which a pestilence flourishes unchallenged. On a scale that will make old wars and past misfortunes look like controlled laboratory experiments. Already 50 million people have been fed into the Development Mill and have emerged as air-conditioners and popcorn and rayon suits – *subsidized* air-conditioners and popcorn and rayon suits. If we must have these nice things – and they *are* nice – at least we should be made to pay for them.

There's a hole in the flag that needs mending.

It's a sad thing to have to say, but as long as we have faith – we have no hope. To hope, we have to *break* the faith. We have to fight specific wars in specific ways and we have to fight to win. Listen then, to the story of the Narmada Valley. Understand it. And, if you wish, enlist. Who knows, it may lead to magic.

*

The Narmada wells up on the plateau of Amarkantak in the Shahdol district of Madhya Pradesh, then winds its way through 1,300 kilometres of beautiful broadleaved forest and perhaps the most fertile agricultural land in India. Twenty-five million people live in the river valley, linked to the ecosystem and to each other by an ancient, intricate web of interdependence (and, no doubt, exploitation).

Though the Narmada has been targeted for 'water resource development' for more than fifty years now, the reason it has, until recently, evaded being captured and dismembered is that it flows through three states – Madhya Pradesh, Maharashtra and Gujarat.

Ninety per cent of the river flows through Madhya Pradesh; it merely skirts the northern border of Maharashtra, then flows through Gujarat for about 180 kilometres before emptying into the Arabian sea at Bharuch.

As early as 1946, plans had been afoot to dam the river at Gora in Gujarat. In 1961, Nehru laid the foundation stone for a 49.8 metre-high dam – the midget progenitor of the Sardar Sarovar.

Around the same time, the Survey of India drew up new topographical maps of the river basin. The dam planners in Gujarat studied the new maps and decided that it would be more profitable to build a much bigger dam. But this meant hammering out an agreement with neighbouring states.

For years the three states bickered and balked but failed to agree on a water-sharing formula. Eventually, in 1969, the Central Government set up the Narmada Water Disputes Tribunal. It took the Tribunal another ten years to announce its Award. *The people whose lives were going to be devastated were neither informed nor consulted nor heard.*

To apportion shares in the waters, the first, most basic thing the Tribunal had to do was to find out how much water there was in the river. Usually this can only be reliably estimated if there are at least forty years of recorded data on the volume of actual flow in the river. Since this was not available, they decided to extrapolate from rainfall data. They arrived at a figure of 27.22 million acre feet (MAF).[24]

This figure is the statistical bedrock of the Narmada Valley Projects. We are still living with its legacy. It more or less determines the overall design of the Projects – the height, location and number of dams. By inference, it determines the cost of the Projects, how much area will be submerged, how many people will be displaced and what the benefits will be.

In 1992, actual observed flow data for the Narmada – which was now available for forty-five years (from 1948 to 1992) – showed that the yield from the river was only 22.69 MAF – 18 per cent less![25] The Central Water Commission admits that there is less water in the Narmada than had previously been assumed.[26] The Government of India says:

> *It may be noted that clause II* (of the decision of the Tribunal) *relating to determination of dependable flow as 28 MAF is non-reviewable* (!)[27]

Never mind the data – the Narmada is legally bound by human decree to produce as much water as the Government of India commands.

Its proponents boast that the Narmada Valley Project is the most ambitious river valley project ever conceived in human history. They plan to build 3,200 dams that will reconstitute the Narmada and her forty-one tributaries into a series of step reservoirs – an immense staircase of amenable water. Of these, thirty will be major dams, 135 medium and the rest small. Two of the major dams will be multi-purpose mega-dams. The Sardar Sarovar in Gujarat and the Narmada Sagar in Madhya Pradesh will, between them, hold more water than any other reservoir on the Indian subcontinent.

Whichever way you look at it, the Narmada Valley Development Project is Big. It will alter the ecology of the entire river basin of one of India's biggest rivers. For better or for worse, it will affect the lives of 25 million people who live in the valley. It will submerge and destroy 4,000 square kilometres of natural deciduous forest.[28] Yet, even before the Ministry of Environment cleared the project, the

33

World Bank offered to finance the linchpin of the project – the Sardar Sarovar dam, whose reservoir displaces people in Madhya Pradesh and Maharashtra, but whose benefits go to Gujarat. The Bank was ready with its cheque-book *before* any costs were computed, *before* any studies had been done, *before* anybody had any idea of what the human cost or the environmental impact of the dam would be!

The $450 million loan for the Sardar Sarovar Projects was sanctioned and in place in 1985. The Ministry of Environment clearance for the project came only in 1987! Talk about enthusiasm. It fairly borders on evangelism. Can anybody care so much?

Why were they so keen?

Between 1947 and 1994 the World Bank's management submitted 6,000 projects to the Executive Board. The board hasn't turned down a single one. *Not a single one*. Terms like 'Moving money' and 'Meeting loan targets' suddenly begin to make sense.

India is in a situation today where it pays back more money to The Bank in interest and repayment

instalments than it receives from it. We are forced
to incur new debts in order to be able to repay
our old ones. According to the *World Bank Annual
Report*, last year (1998), after the arithmetic, India
paid The Bank $478 million more than it borrowed.
Over the last five years (1993 to 1998) India paid
The Bank $1.475 billion more than it received.[29]
The relationship between us is exactly like the
relationship between a landless labourer steeped in
debt and the village moneylender – it is an affec-
tionate relationship, the poor man loves his
moneylender because he's always there when he's
needed. It's not for nothing that we call the world
a Global Village. The only difference between the
landless labourer and the Government of India is
that one uses the money to survive. The other just
funnels it into the private coffers of its officers and
agents, pushing the country into an economic
bondage that it may never overcome.

The international Dam Industry is worth $20
billion a year.[30] If you follow the trails of Big Dams
the world over, wherever you go – China, Japan,
Malaysia, Thailand, Brazil, Guatemala – you'll rub
up against the same story, encounter the same
actors: the Iron Triangle (dam-jargon for the nexus

between politicians, bureaucrats and dam construction companies), the racketeers who call themselves International Environmental Consultants (who are usually directly employed by dam-builders or their subsidiaries), and more often than not, the friendly neighbourhood World Bank. You'll grow to recognise the same inflated rhetoric, the same noble 'People's Dam' slogans, the same swift, brutal repression that follows the first sign of civil insubordination. (Of late, especially after its experience in the Narmada Valley, The Bank is more cautious about choosing the countries in which it finances projects that involve mass displacement. At present, China is its Most Favoured client. It's the great irony of our times – American citizens protest the massacre in Tiananmen Square, but The Bank has used their money to fund studies for the Three Gorges dam in China which is going to displace 1.3 million people. The Bank is today the biggest foreign financier of large dams in China.[31])

It's a skilful circus and the acrobats know each other well. Occasionally they'll swap parts – a bureaucrat will join The Bank, a Banker will surface as a Project Consultant. At the end of play, a huge percentage of what's called 'Development Aid' is rechannelled back to the countries it came from, masquerading as equipment cost or consultants' fees or salaries to the agencies' own staff. Often Aid is openly 'tied' (as in the case of the Japanese loan for the Sardar Sarovar dam – to a contract for purchasing turbines from the Sumitomo Corporation).[32] Sometimes the connections are more murky. In 1993, Britain financed the Pergau Dam in Malaysia with a subsidised loan of £234 million, despite an Overseas Development Administration report that said that the dam would be a 'bad buy' for Malaysia. It later emerged that the loan was offered to 'encourage' Malaysia to sign a £1.3 *billion* contract to buy British arms.[33]

In 1994, British consultants earned $2.5 billion on overseas contracts.[34] The second biggest sector of the market after Project Management was writing what are called EIAs (Environmental Impact Assessments). In the Development racket, the rules are pretty simple. If you get invited by a Government to write an EIA for a big dam project and you point out a problem (say, you quibble about the amount of water available in a river, or, God forbid, you suggest that the human costs are perhaps too high) then you're history. You're an OOWC. An Out Of

Work Consultant. And oops! There goes your Range Rover. There goes your holiday in Tuscany. There goes your children's private boarding school. There's good money in poverty. Plus Perks.

In keeping with Big Dam tradition, concurrent with the construction of the 138.68 metre-high Sardar Sarovar dam, began the elaborate Government pantomime of conducting studies to estimate the actual project costs and the impact it would have on people and the environment. The World Bank participated wholeheartedly in the charade – occasionally it beetled its brows and raised feeble requests for more information on issues like the resettlement and rehabilitation of what it calls PAPs – Project Affected Persons. (They help, these acronyms, they manage to mutate muscle and blood into cold statistics. PAPs soon cease to be people). The merest crumbs of information satisfied The Bank and it proceeded with the project. The implicit, unwritten but fairly obvious understanding between the concerned agencies was that whatever the costs – economic, environmental or human – the project would go ahead. They would justify it as they went along. They knew full well that eventually, in a courtroom or to a committee,

no argument works as well as a fait accompli.
M' lord, the country is losing two crores a day due to the delay.

The Government refers to the Sardar Sarovar Projects as the 'Most Studied Project in India', yet the game goes something like this: when the Tribunal first announced its Award and the Gujarat Government announced its plan of how it was going to use its share of water, *there was no mention of drinking water for villages in Kutch and Saurashtra*, the arid areas of Gujarat. When the project ran into political trouble, the Government suddenly discovered the emotive power of Thirst. Suddenly, quenching the thirst of parched throats in Kutch and Saurashtra became the whole *point* of the Sardar Sarovar Projects. (Never mind that water from two rivers – the Sabarmati and the Mahi, both of which are *miles* closer to Kutch and Saurashtra than the Narmada, have been dammed and diverted to Ahmedabad, Mehsana and Kheda. Neither Kutch nor Saurashtra has seen a drop of it.) Officially, the number of people who will be provided drinking water by the Sardar Sarovar Canal fluctuates from 28 million (1983) to 32.5 million (1989) – nice touch, the decimal point! – to

10 million (1992) and down to 25 million (1993).[35] In 1979 the number of villages that would receive drinking water was zero. In the early 80s it was 4,719, in 1990 it was 7,234 and in 1991 it was 8,215.[36] When pressed, the Government admitted that the figures for 1991 included 236 *uninhabited* villages![37]

Every aspect of the project is approached in this almost playful manner, as if it's a family board game. Even when it concerns the lives and futures of vast numbers of people.

In 1979 the number of families that would be displaced by the Sardar Sarovar reservoir was estimated to be a little over 6,000. In 1987 it grew to 12,000. In 1991 it surged to 27,000. In 1992 the Government acknowledged that 40,000 families would be affected. Today, the official figure hovers between 40,000 and 41,500.[38] (Of course even this is an absurd figure, because the reservoir isn't the *only* thing that displaces people. According to the NBA the actual figure is about 85,000 families – that's *half a million* people.)

The estimated cost of the project bounced up from under Rs 5,000 crores[39] to Rs 20,000 crores (officially). The NBA says that it will cost Rs 44,000 crores.[40]

The Government claims the Sardar Sarovar Projects will produce 1,450 megawatts of power.[41] The thing about multi-purpose dams like the Sardar Sarovar is that their 'purposes' (irrigation, power production and flood-control) conflict with one another. Irrigation uses up the water you need to produce power. Flood control requires you to keep the reservoir empty during the monsoon months to deal with an anticipated surfeit of water. And if there's no surfeit, you're left with an empty dam. And this defeats the purpose of irrigation, which is to *store* the monsoon water. It's like the conundrum of trying to ford a river with a fox, a chicken and a bag of grain. The result of these mutually conflicting aims, studies say, is that when the Sardar Sarovar Projects are completed and the scheme is fully functional, it will end up producing only 3 per cent of the power that its planners say it will. About 50 megawatts. And if you take into account the power needed to pump water through its vast network of canals, the Sardar Sarovar Projects will end up *consuming* more electricity than they produce![42]

In an old war, everybody has an axe to grind. So how do you pick your way through these claims and counter-claims? How do you decide whose estimate is more reliable? One way is to take a look at the track record of Indian dams.

The Bargi dam near Jabalpur was the first dam on the Narmada to be completed (in 1990). It cost ten times more than was budgeted and submerged three times more land than the engineers said it would. About 70,000 people from 101 villages were supposed to be displaced, but when they filled the reservoir (without warning anybody), 162 villages were submerged. Some of the resettlement sites built by the Government were submerged as well. People were flushed out like rats from the land they had lived on for centuries. They salvaged what they could, and watched their houses being washed away. 114,000 people were displaced.[43] There was no rehabilitation policy. Some were given meagre cash compensations. Many got absolutely nothing. A few were moved to government rehabilitation sites. The site at Gorakhpur is, according to Government publicity, an 'ideal village'. Between 1990 and 1992, five people died of starvation there. The rest either returned to live illegally in the forests near the reservoir, or moved to slums in Jabalpur.

The Bargi dam irrigates only as much land as it submerged in the first place – *and only 5 per cent of the area that its planners claimed it would irrigate.*[44] Even that is waterlogged.

Time and again, it's the same story. The Andhra Pradesh Irrigation II scheme claimed it would displace 63,000 people. When completed, it displaced 150,000 people[45]. The Gujarat Medium Irrigation II scheme displaced 140,000 people instead of 63,600.[46] The revised estimate of the number of people to be displaced by the Upper Krishna irrigation project in Karnataka is 240,000 against its initial claims of displacing only 20,000.[47]

These are World Bank figures. Not the NBA's. Imagine what this does to our conservative estimate of 33 million.

Construction work on the Sardar Sarovar dam site, which had continued sporadically since 1961, began in earnest in 1988. At the time, nobody, not the

Government, nor the World Bank, were aware that a woman called Medha Patkar had been wandering through the villages slated to be submerged, asking people whether they had any idea of the plans that the Government had in store for them. When she arrived in the valley all those years ago, opposing the construction of the dam was the furthest thing from her mind. Her chief concern was that displaced villagers should be resettled in an equitable, humane way. It gradually became clear to her that the Government's intentions towards them were far from honourable. By 1986 word had spread and each state had a peoples' organisation that questioned the promises about resettlement and rehabilitation that were being bandied about by Government officials. It was only some years later that the full extent of the horror – the impact that the dams would have, both on the people who were to be displaced and the people who were supposed to benefit – began to surface. The Narmada Valley Development Project came to be known as India's Greatest Planned Environmental Disaster. The various peoples' organisations massed into a single organisation and the Narmada Bachao Andolan – the extraordinary NBA – was born.

In 1988 the NBA formally called for all work on the Narmada Valley Development Projects to be stopped. People declared that they would drown if they had to, but would not move from their homes. Within two years, the struggle had burgeoned and had support from other resistance movements. In September 1989, more than 50,000 people gathered in the Valley from all over India to pledge to fight Destructive Development. The dam site and its adjacent areas, already under the Indian Official Secrets Act, was clamped under Section 144 which prohibits the gathering of groups of more than five people. The whole area was turned into a police camp. Despite the barricades, one year later, on 28 September 1990, thousands of villagers made their way on foot and by boat to a little town called Badwani, in Madhya Pradesh, to reiterate their pledge to drown rather than agree to move from their homes.

News of the people's opposition to the Projects spread to other counties. The Japanese arm of Friends of the Earth mounted a campaign in Japan that succeeded in getting the Government of Japan to withdraw its 27 billion yen loan to finance the Sardar Sarovar Projects. (The contract for the turbines still holds.) Once the Japanese withdrew, international pressure from various environmental activist groups who supported the struggle began to mount on the World Bank.

This of course, led to an escalation of repression in the valley. Government policy, described by a particularly articulate minister, was to 'flood the valley with khaki'.

On Christmas Day in 1990, six thousand men and women walked over a hundred kilometres, carrying their provisions and their bedding, accompanying a seven-member sacrificial squad that had resolved to lay down its lives for the river. They were stopped at Ferkuwa on the Gujarat border by battalions of armed police and crowds of people from the city of Baroda, many of whom were hired, some of whom perhaps genuinely believed that the Sardar Sarovar was 'Gujarat's life-line'. It was a telling confrontation. Middle Class Urban India vs. a Rural, predominantly Adivasi, Army. The marching people demanded they be allowed to cross the border and walk to the dam site. The police refused them passage. To stress their commitment to non-

46

violence, each villager had his or her hands bound together. One by one, they defied the battalions of police. They were beaten, arrested and dragged into waiting trucks in which they were driven off and dumped some miles away, in the wilderness. They just walked back and began all over again.

The face-off continued for almost two weeks. Finally, on 7 January 1991, the seven members of the sacrificial squad announced that they were going on an indefinite hunger strike. Tension rose to dangerous levels. The Indian and international Press, TV camera crews and documentary filmmakers, were present in force. Reports appeared in the papers almost every day. Environmental Activists stepped up the pressure in Washington. Eventually, acutely embarrassed by the glare of unfavourable media, the World Bank announced that it would commission an independent review of the Sardar Sarovar Projects – unprecedented in the history of Bank behaviour. When the news reached the valley, it was received with distrust and uncertainty. The people had no reason to trust the World Bank. But still, it was a victory of sorts. The villagers, understandably upset by the frightening deterioration in the condition of their comrades who had not eaten for twenty-two days, pleaded with them to call off the fast. On 28 January, the fast at Ferkuwa was called off and the brave, ragged army returned to their homes shouting. '*Hamara gaon mein hamara Raj!*' (Our Rule in our villages).

There has been no army quite like this one, any-where else in the world. In other countries – China (Chairman Mao got a Big Dam for his seventy-seventh birthday), Malaysia, Guatemala, Paraguay – every sign of revolt has been snuffed out almost before it began. Here in India, it goes on and on. Of course, the State would like to take credit for this too. It would like us to be grateful to it for not crushing the movement completely, for *allowing* it to exist. After all what *is* all this, if not a sign of a healthy functioning democracy in which the State has to intervene when its people have differences of opinion?

I suppose that's one way of looking at it. (Is this my cue to cringe and say 'Thank you, thank you, for allowing me to write the things I write?')

We don't need to be grateful to the State for per-mitting us to protest. We can thank ourselves for that. It is we who have insisted on these rights. It is we who have refused to surrender them. If we have anything to be truly proud of as a people, it is this.

The struggle in the Narmada valley lives, *despite* the State.

The Indian State makes war in devious ways. Apart from its apparent benevolence, its other big weapon is its ability to wait. To roll with the punches. To wear out the opposition. The State never tires, never ages, never needs a rest. It runs an endless relay.

But fighting people tire. They fall ill, they grow old. Even the young age prematurely. For twenty years now, since the Tribunal's award, the ragged army in the valley has lived with the fear of eviction. For twenty years, in most areas there has been no sign of 'development' – no roads, no schools, no wells, no medical help. For twenty years, it has borne the stigma 'slated for submergence' – so it's isolated from the rest of society (no marriage proposals, no land transactions). They're a bit like the Hibakushas in Japan (the victims and their descendants of the bombing in Hiroshima and Nagasaki). The 'fruits of modern development', when they finally came, brought only horror. Roads brought surveyors. Surveyors brought trucks. Trucks brought policemen. Policemen brought bullets and beatings and rape and arrest and in one case, murder. The only genuine 'fruit' of modern development that reached them, reached them inadvertently – the right to raise their voices, the right to be heard. But they have fought for twenty years now. How much longer will they last?

The struggle in the valley is tiring. It's no longer as fashionable as it used to be. The international camera crews and the radical reporters have moved (like the World Bank) to newer pastures. The documentary films have been screened and appreciated. Everybody's sympathy is all used up. But the dam goes on. It's getting higher and higher . . .

Now, more than ever before, the ragged army needs reinforcements. If we let it die, if we allow the struggle to be crushed, if we allow the people to be brutalized, we will lose the most precious thing we have: our spirit, or what's left of it.

'India will go on,' they'll tell you, the sage philosophers who don't want to be troubled by piddling Current Affairs. As though 'India' is somehow more valuable than her people.

Old Nazis probably soothe themselves in similar ways. It's too late, some people say. Too much time and money has gone into the project to revoke it now.

So far, the Sardar Sarovar reservoir has submerged only a fourth of the area that it will when (if) the dam reaches its full height. If we stop it now, we would save 325,000 people from certain destitution. As for the economics of it – it's true that the Government has already spent Rs 7,500 crores, but continuing with the project would mean throwing good money after bad. We would save something like Rs 35,000 crores of public money, enough to fund local water harvesting projects in every village in this vast country. What could possibly be a more worthwhile war?

The war for the Narmada Valley is not just some exotic tribal war, or a remote rural war or even an exclusively Indian war. It's a war for the rivers and the mountains and the forests of the world. All sorts of warriors from all over the world, anyone who wishes to enlist, will be honoured and welcomed. Every kind of warrior will be needed. Doctors, lawyers, teachers, judges, journalists, students, sportsmen, painters, actors, singers, lovers . . . The borders are open, folks! Come on in.

*

Anyway, back to the story.

In June 1991, the World Bank appointed Bradford Morse, a former head of the United Nations Development Program, as Chairman of the Independent Review. His brief was to make a thorough assessment of the Sardar Sarovar Projects. He was guaranteed free access to all secret Bank documents relating to the Projects.

Morse and his team arrived in India in September 1991. The NBA, convinced that this was yet another set-up, at first refused to meet them. The Gujarat Government welcomed the team with a red carpet (and a nod and a wink) as covert allies.

A year later, in June 1992, the historic Independent Review (known also as the Morse Report) was published.

The Independent Review unpeels the project delicately, layer by layer, like an onion. Nothing was too big, and nothing too small for the members of the Morse Committee to enquire into. They met ministers and bureaucrats, they met NGOs working in the area, went from village to village, from resettlement site to resettlement site. They visited the good ones. The bad ones. The temporary ones, the permanent ones. They spoke to hundreds of people. They travelled extensively in the submergence area and the command area. They went to Kutch and other drought-hit areas in Gujarat. They commissioned their own studies. They examined every aspect of the project: hydrology and water management, the upstream environment, sedimentation, catchment area treatment, the downstream environment, the anticipation of likely problems in the command area – waterlogging, salinity, drainage, health, the impact on wildlife.

What the Independent Review reveals, in temperate, measured tones (which I admire, but cannot achieve) is scandalous. It is the most balanced, unbiased, yet damning indictment of the relationship between the Indian State and the World Bank. Without appearing to, perhaps even without intending to, the report cuts through to the cosy core, to the space where they live together and love each other (somewhere between what they say and what they do).

They core recommendation of the 357-page Independent Review was unequivocal and wholly unexpected:

> We think the Sardar Sarovar Projects as they stand are flawed, that resettlement and rehabilitation of all those displaced by the Projects is not possible under prevailing circumstances, and that environmental impacts of the Projects have not been properly considered or adequately addressed. Moreover we believe that the Bank shares responsibility with the borrower for the situation that has developed ... it seems clear that engineering and economic imperatives have driven the Projects to the exclusion of human and environmental concerns ... India and the states involved ... have spent a great deal of

money. No one wants to see this money wasted. But we caution that it may be more wasteful to proceed without full knowledge of the human and environmental costs ... As a result, we think that the wisest course would be for the Bank to step back from the Projects and consider them afresh ...[48]

Four committed, knowledgeable, truly independent men – they do a lot to make up for the faith eroded by hundreds of other venal ones who are paid to do similar jobs.

The World Bank, however, was still not prepared to give up. It continued to fund the project. Two months after the Independent Review, it sent out the Pamela Cox Committee which did exactly what the Morse Review had cautioned against ('... *it would be irresponsible for us to patch together a series of recommendations on implementation when the flaws in the Projects are as obvious as they seem to us ...*'[49]) and suggested a sort of patchwork remedy to try and salvage the operation. In October 1992, on the recommendation of the Pamela Cox Committee, the Bank asked the Indian Government to meet some minimum, primary conditions within a period of six months.[50] Even that much, the Government couldn't do. Finally, on 30 March 1993, the World Bank pulled out of the Sardar Sarovar Projects. (Actually, technically, on 29 March, one day *before* the deadline, the Government of India asked the World Bank to withdraw.)[51] Details. Details.

No-one has ever managed to make the World Bank step back from a project before. Least of all a rag-tag army of the poorest people in one of the world's poorest countries. A group of people whom Lewis Preston, then President of The Bank, never managed to fit into his busy schedule when he visited India.[52] Sacking The Bank was and is a huge moral victory for the people in the valley.

The euphoria didn't last. The Government of Gujarat announced that it was going to raise the $200 million shortfall on its own and push ahead with the project.

During the period of the Independent Review and after it was published, confrontation between people and the Authorities continued unabated in the valley – humiliation, arrests, baton charges. Indefinite fasts terminated by temporary promises and permanent betrayals. People who had agreed to leave the valley and be resettled had begun returning to their villages from their resettlement sites. In Manibeli, a village in Maharashtra and one of the nerve-centres of the resistance, hundreds of villagers participated in a Monsoon Satyagraha. In 1993, families in Manibeli remained in their homes as the waters rose. They clung to wooden posts with their children in their arms and refused to move. Eventually policemen prised them loose and dragged them away. The NBA declared that if the Government did not agree to review the project, on 6 August 1993 a band of activists would drown themselves in the rising waters of the reservoir. On 5 August, the Union Government constituted yet another committee called the Five Member Group (FMG) to review the Sardar Sarovar Projects. The Government of Gujarat refused it entry into Gujarat.[53]

The FMG report[54] (a 'desk report') was submitted the following year. It tacitly endorsed the grave concerns of the Independent Review. But it made no difference. Nothing changed. This is another of the State's tested strategies. It kills you with committees.

In February 1994, the Government of Gujarat ordered the permanent closure of the sluice gates of the dam.
In May 1994, the NBA filed a writ petition in the Supreme Court questioning the whole basis of the Sardar Sarovar dam and seeking a stay on its construction.[55]

That monsoon, when the level in the reservoir rose and the water smashed down on the other side of the dam, 65,000 cubic metres of concrete and 35,000 cubic metres of rock were torn out of a stilling basin, leaving a crater sixty-five metres wide. The riverbed powerhouse was flooded. The damage was kept secret for months.[56] Reports started appearing about it in the press only in January of 1995.

In early 1995, on the grounds that the rehabilitation of displaced people had not been adequate,

the Supreme Court ordered work on the dam
to be suspended until further notice.[57] The height
of the dam was eighty metres above mean sea
level.

Meanwhile, work had begun on two more dams
in Madhya Pradesh – the massive Narmada Sagar
(without which the Sardar Sarovar loses 17 per cent
to 30 per cent of its efficiency[58]) and the Maheshwar
dam. The Maheshwar dam is next in line, upstream
from the Sardar Sarovar. The Government of
Madhya Pradesh has signed a power purchase contract with a private company – S. Kumars – one of
India's leading textile magnates.

Tension in the Sardar Sarovar area abated tempor-
arily and the battle moved upstream, to Maheshwar, in the fertile plains of Nimad.

The case pending in the Supreme Court led to a
palpable easing of repression in the valley. Construction work had stopped on the dam, but the
rehabilitation charade continued. Forests (slated for
submergence) continued to be cut and carted away

in trucks, forcing people who depended on them for a livelihood to move out.

Even though the dam is nowhere near its eventual projected height, its impact on the environment and the people living along the river is already severe.

Around the dam site and the nearby villages, the number of cases of malaria has increased six-fold.[59]

Several kilometres upstream from the Sardar Sarovar dam, huge deposits of silt, hip-deep and over two hundred metres wide, have cut off access to the river. Women carrying water pots now have to walk miles, literally *miles*, to find a negotiable entry point. Cows and goats get stranded in the mud and die. The little single-log boats that Adivasis use, have become unsafe on the irrational circular currents caused by the barricade downstream.

Further upstream, where the silt deposits have not yet become a problem, there's another tragedy. Landless people (predominantly Adivasis and Dalits), have traditionally cultivated rice, melons, cucumbers and gourds on the rich, shallow silt banks the river leaves when it recedes in the dry months. Every now and then, the engineers manning the Bargi dam (way upstream, near Jabalpur) release water from the reservoir without warning. Downstream, the water level in the river suddenly rises. Hundreds of families have had their crops washed away several times, leaving them with no livelihood.

Suddenly they can't trust their river any more. It's like a loved one who has developed symptoms of psychosis. Anyone who has loved a river can tell you that the loss of a river is a terrible, aching thing. But I'll be rapped on the knuckles if I continue in this vein. When we're discussing the Greater Common Good there's no place for sentiment. One must stick to facts. Forgive me for letting my heart wander.

*

The State Governments of Madhya Pradesh and Maharashtra continue to be completely cavalier in their dealings with displaced people. The Government of Gujarat has a rehabilitation policy (on paper) that makes the other two states look medieval. It boasts of being the best rehabilitation package in the world.[60] It offers land for land to displaced people from Maharashtra and Madhya Pradesh and recognises the claims of 'encroachers' (usually Adivasis with no papers). The deception however, lies in its definition of who qualifies as 'Project Affected'.

In point of fact, the Government of Gujarat hasn't even managed to rehabilitate people from its own nineteen villages slated for submergence, let alone the rest of the 226 villages in the other two states. The inhabitants of these nineteen villages have been scattered to 175 separate rehabilitation sites. Social links have been smashed, communities broken up.

In practice, the resettlement story (with a few 'Ideal Village' exceptions) continues to be one of callousness and broken promises. Some people have been given land, others haven't. Some have land that is stony and uncultivable. Some have land that is irredeemably waterlogged. Some have been driven out by landowners who had sold their land to the Government but hadn't been paid.[61]

Some who were resettled on the periphery of other villages have been robbed, beaten and chased away by their host villagers. There have been instances when displaced people from two different dam projects have been allotted contiguous lands. In one case, displaced people from *three* dams – the Ukai dam, the Sardar Sarovar dam and the Karjan dam – were resettled in the *same* area.[62] In addition to fighting amongst themselves for resources – water, grazing land, jobs – they had to fight a group of landless labourers who had been sharecropping the land for absentee landlords who had subsequently sold it to the Government.

There's another category of displaced people – people whose lands have been acquired by the Government for resettlement sites. There's a pecking order even amongst the wretched – Sardar Sarovar 'oustees' are more glamorous than other 'oustees' because they're occasionally in the news and have a case in court. (In other Development Projects where there's no press, no NBA, no court case, there are no records. The displaced leave no trail at all.)

In several resettlement sites, people have been dumped in rows of corrugated tin sheds which are furnaces in summer and fridges in winter. Some of them are located in dry river beds that, during the monsoon, turn into fast-flowing drifts. I've been to some of these 'sites'. I've seen film footage[63] of others: shivering children, perched like birds on the edges of charpais, while swirling waters enter their tin homes. Frightened, fevered eyes watch pots and pans carried through the doorway by the current, floating out into the flooded fields, thin fathers swimming after them to retrieve what they can.

When the waters recede they leave ruin. Malaria, diarrhoea, sick cattle stranded in the slush. The ancient teak beams dismantled from their previous homes, carefully stacked away like postponed dreams, now spongy, rotten and unusable.

Forty households were moved from Manibeli to a resettlement site in Gujarat. In the first year, thirty-eight children died.[64] In today's papers (*Indian Express*, 26 April '99) there's a report about nine deaths in a single rehabilitation site in Gujarat.

In the course of a single week. That's 1.2875 PAPs a day, if you're counting.

Many of those who have been resettled are people who have lived all their lives deep in the forest with virtually no contact with money and the modern world. Suddenly they find themselves left with the option of starving to death or walking several kilo-metres to the nearest town, sitting in the market-place (both men and women) offering themselves as wage labour, like goods on sale.

Instead of a forest from which they gathered every-thing they needed – food, fuel, fodder, rope, gum, tobacco, tooth powder, medicinal herbs, housing materials – they earn between ten and twenty rupees a day with which to feed and keep their families. Instead of a river, they have a hand pump. In their old villages, they had no money, but they were insured. If the rains failed, they had the forests to turn to. The river to fish in. Their livestock was their fixed deposit. Without all this, they're a heartbeat away from destitution.

In Vadaj, a resettlement site I visited near Baroda, the man who was talking to me rocked his sick baby in his arms, clumps of flies gathered on its sleeping eyelids. Children collected around us, taking care not to burn their bare skin on the scorching tin walls of the shed they call a home. The man's mind was far away from the troubles of his sick baby. He was making me a list of the fruit he used to pick in the forest. He counted forty-eight kinds. He told me that he didn't think he or his children would ever be able to afford to eat any fruit again. Not unless he stole it. I asked him what was wrong with his baby. He said it would be better for the baby to die than live like this. I asked what the baby's mother thought about that. She didn't reply. She just stared.

For the people who've been resettled, everything has to be relearned. Every little thing, every big thing: from shitting and pissing (where d'you do it when there's no jungle to hide you?) to buying a bus ticket, to learning a new language, to understanding money. And worst of all, learning to be supplicants. Learning to take orders. Learning to have Masters. Learning to answer only when you're addressed. In addition to all this, they have to learn how to

make written representations (in triplicate) to the
Grievance Redressal Committee or the Sardar
Sarovar Narmada Nigam for any particular problems they might have. Recently, 3,000 people came
to Delhi to protest their situation – travelling overnight by train, living on the blazing streets.[65] The
President wouldn't meet them because he had an
eye infection. Maneka Gandhi, the Minister for
Social Justice and Empowerment, wouldn't meet
them but asked for a written representation (*Dear
Maneka, Please don't build the dam, Love, The People*).
When the representation was handed to her she
scolded the little delegation for not having written
it in English.

From being self-sufficient and free, to being impoverished and yoked to the whims of a world you
know nothing, *nothing* about – what d'you suppose
it must feel like? Would you like to trade your
beach house in Goa for a hovel in Paharganj? No?
Not even for the sake of the Nation?

Truly, it is just not possible for a State Administration, *any* State Administration, to carry out the
rehabilitation of a people as fragile as this, on such

an immense scale. It's like using a pair of hedge-shears to trim an infant's fingernails. You can't do it without shearing its fingers off.

Land for land sounds like a reasonable swap, but how do you implement it? How do you uproot 200,000 people (the official blinkered estimate) – of whom 117,000 are Adivasi – and relocate them in a humane fashion? How do you keep their communities intact in a country where every inch of land is fought over, where almost all litigation pending in courts has to do with land disputes? Where is all this fine, unoccupied but arable land that is waiting to receive these intact communities?

The simple answer is that there isn't any. Not even for the 'officially' displaced of this one dam.

What about the rest of the 3,199 dams? What about the remaining thousands of 'PAPs' earmarked for annihilation? Shall we just put the Star of David on their doors and get it over with?

*

The reservoir of the Maheshwar dam will wholly or partially submerge sixty villages in the Nimad plains of Madhya Pradesh. A significant section of the population in these villages – roughly a third – are Kevats and Kahars, ancient communities of ferrymen, fisherfolk, sand quarriers and cultivators of the riverbank when the waters recede in the dry season. Most of them own no land, but the river sustains them, and means more to them than to anyone else. When the dam is built, thousands of Kevats and Kahars will lose their only source of livelihood. Yet, simply because they are landless, they do not qualify as Project-affected and will not be eligible for rehabilitation.

Jalud is the first of sixty villages that will be submerged by the reservoir of the Maheshwar dam. Jalud is not an Adivasi village, and is therefore riven with the shameful caste divisions that are the scourge of every ordinary Hindu village. A majority of the land-owning farmers (the ones who qualify as PAPs) are Rajputs. They farm some of the most fertile soil in India. Their houses are piled with sacks of wheat and daal and rice. They boast so much about the things they grow on their land that if it weren't so tragic, it could get on your nerves. Their houses have already begun to crack with the impact of the dynamiting on the dam site.

Twelve families who had smallholdings in the vicinity of the dam site had their land acquired. They told me how, when they objected, cement was poured into their water pipes, their standing crops were bulldozed and the police occupied the land by force. All twelve families are now landless and work as wage labourers.

The area that the Rajputs of Jalud are going to be moved to is a few kilometres inland, away from the river, adjoining a predominantly Dalit and Adivasi precinct in a village called Samraj. I saw the huge tract of land that had been marked off for them. It was a hard, stony hillock with stubbly grass and scrub, on which truckloads of silt were being unloaded and spread out in a thin layer to make it look like rich, black humus.

The story goes like this: on behalf of the S. Kumars (Textile Tycoons turned Nation Builders) the District Magistrate acquired the hillock, which was actually village common grazing land that belonged to the people of Samraj. In addition to this, the land of thirty-four Dalit and Adivasi villagers was acquired. No compensation was paid.

The villagers, whose main source of income was their livestock, had to sell their goats and buffaloes because they no longer had anywhere to graze them. Their only remaining source of income lies (lay) on the banks of a small lake on the edge of the village. In summer, when the water level recedes, it leaves a shallow ring of rich silt on which the villagers grow (grew) rice and melons and cucumber. The S. Kumars have excavated this silt, to cosmetically cover the stony grazing ground (that the Rajputs of Jalud don't want). The banks of the lake are now steep and uncultivable.

The already impoverished people of Samraj have been left to starve, while this photo opportunity is being readied for German and Swiss funders, Indian courts and anybody else who cares to pass that way.

*

This is how India works. This is the genesis of the Maheshwar dam. The story of the first village. What will happen to the other fifty-nine? May bad luck pursue this dam. May bulldozers turn upon the Textile Tycoons.

Nothing can justify this kind of behaviour.

In circumstances like these, to even entertain a debate about Rehabilitation is to take the first step towards setting aside the Principles of Justice. Resettling 200,000 people in order to take (or pretend to take) drinking water to 40 million – there's something very wrong with the *scale* of operations here. This is Fascist Maths. It strangles stories. Bludgeons detail. And manages to blind perfectly reasonable people with its spurious, shining vision.

*

When I arrived on the banks of the Narmada in late March 1999, it was a month after the Supreme Court had suddenly vacated the stay on construction work of the Sardar Sarovar dam. I had read pretty much everything I could lay my hands on (all those 'secret' Government documents). I had a clear idea of the lay of the land – of what had happened where and when and to whom. The story played itself out before my eyes like a tragic film whose actors I'd already met. Had I not known its history, nothing would have made sense. Because in the valley there are stories within stories and it's easy to lose the clarity of rage in the sludge of other people's sorrow.

I ended my journey in Kevadia Colony, where it all began.

Thirty-eight years ago, this is where the Government of Gujarat decided to locate the infrastructure it would need for starting work on the dam: guest houses, office blocks, accommodation for engineers and their staff, roads leading to the dam site, warehouses for construction material.

It is located on the cusp of what is now the Sardar Sarovar reservoir and the Wonder Canal, Gujarat's 'lifeline', that is going to quench the thirst of millions.

Nobody knows this, but Kevadia Colony is the key to the World. Go there, and secrets will be revealed to you.

*

In the winter of 1961, a government officer arrived in a village called Kothie and told the villagers that some of their land would be needed to construct a helipad because someone terribly important was going to come visiting. In a few days, a bulldozer arrived and flattened standing crops. The villagers were made to sign papers and were paid a sum of money, which they assumed was payment for their destroyed crops. When the helipad was ready, a helicopter landed on it, and out came Prime Minister Nehru. Most of the villagers couldn't see him because he was surrounded by policemen. Nehru made a speech. Then he pressed a button and there was an explosion on the other side of the river. After the explosion he flew away.[66] That was the genesis of what was to become the Sardar Sarovar dam.

Could Nehru have known when he pressed that button that he had unleashed an incubus?

After Nehru left, the Government of Gujarat arrived in strength. It acquired 1,600 acres of land from 950 families from six villages.[67] The people were Tadvi Adivasis who, because of their proximity to the city of Baroda, were not entirely unversed in the ways of a market economy. They were sent notices and told that they would be paid cash compensations and given jobs on the dam site. Then the nightmare began.

Trucks and bulldozers rolled in. Forests were felled, standing crops destroyed. Everything turned into a whirl of jeeps and engineers and cement and steel. Mohan Bhai Tadvi watched eight acres of his land with standing crops of jowar, toovar and cotton being levelled. Overnight he became a land-less labourer. *Three years later* he received his cash compensation of Rs 250 an acre in three separate instalments.

Dersukh Bhai Vesa Bhai's father was given Rs 3,500 for his house and five acres of land with its standing crops and all the trees on it. He remembers walking all the way to Rajpipla (the district headquarters) as a little boy, holding his father's hand.

He remembers how terrified they were when they were called in to the Tehsildar's office. They were made to surrender their compensation notices and sign a receipt. They were illiterate, so they didn't know how much the receipt was made out for.

Everybody had to go to Rajpipla but they were always summoned on different days, one by one. So they couldn't exchange information or compare stories.

Gradually, out of the dust and bulldozers, an offensive, diffuse configuration emerged. Kevadia Colony. Row upon row of ugly cement flats, offices, guest houses, roads. All the graceless infrastructure of Big Dam construction. The villagers' houses were dismantled and moved to the periphery of the colony where they remain today, squatters on their own land. Those that caused trouble were intimidated by the police and the construction company. The villagers told me that in the contractor's headquarters they have a 'lock-up' like a police lock-up, where recalcitrant villagers are incarcerated and beaten.

The people who were evicted to build Kevadia Colony do not qualify as 'Project-Affected' in Gujarat's Rehabilitation package.

Some of them work as servants in the officers' bungalows and waiters in the guest house built on the land where their own houses once stood. Can there be anything more poignant?

Those who had some land left, tried to cultivate it, but Kevadia municipality introduced a scheme in which they brought in pigs to eat uncollected refuse on the streets. The pigs stray into the villagers' fields and destroy their crops.

In 1992, thirty years later, each family has been offered a sum of Rs 12,000 per acre, up to a maximum of Rs 36,000, *provided* they agree to leave their homes and go away! Yet 40 per cent of the land that was acquired is lying unused. The Government refuses to return it. Eleven acres, acquired from Deviben, who is a widow now, have been given over to the Swami Narayan Trust (a big religious sect). On a small portion of it, the Trust runs a little school. The rest it cultivates, while Deviben watches through the barbed-wire fence. On 200 acres acquired in the village of Gora, villagers were evicted and blocks of flats were built. They lay empty for years. Eventually the Government hired it for a nominal fee to Jai Prakash Associates, the dam contractors, who, the villagers say, sublet it privately for Rs 32,000 a month. (Jai Prakash Associates, the biggest dam contractors in the country, the *real* nation-builders, own the Siddharth Continental and the Vasant Continental Hotels in Delhi.)

On an area of about thirty acres there is an absurd cement Public Works Department replica of the ancient Shoolpaneshwar temple that was submerged in the reservoir. The same political formation that plunged a whole nation into a bloody, medieval nightmare because it insisted on destroying an old mosque to dig up a non-existent temple, thinks nothing of submerging a hallowed pilgrimage route and hundreds of temples that have been worshipped in for centuries.

It thinks nothing of destroying the sacred hills and groves, the places of worship, the ancient homes of the gods and demons of the Adivasi.

It thinks nothing of submerging a valley that has yielded fossils, microliths and rock paintings, the only valley in India, according to archaeologists, that contains an uninterrupted record of human occupation from the Old Stone Age.

What can one say?

In Kevadia Colony, the most barbaric joke of all is the wildlife museum. The Shoolpaneshwar Sanctuary Interpretation Centre gives you quick, comprehensive evidence of the Government's sincere commitment to Conservation.

The Sardar Sarovar reservoir, when the dam reaches its full height, is going to submerge about 13,000 hectares of prime forest land. (In anticipation of submergence, the forest began to be felled many greedy years ago.) Between the Narmada Sagar dam and the Sardar Sarovar dam, 50,000 hectares of old growth broadleaved forest will be submerged. Madhya Pradesh has the highest rate of forest cover loss in the whole of India. This is partly responsible for the reduced flow in the Narmada and the increase in siltation. Have engineers made the connection between forest, rivers and rain? Unlikely. It isn't part of their brief. Environmentalists and conservationists were quite rightly alarmed at the extent of loss of biodiversity and wildlife habitat that the submergence would cause. To mitigate this loss, the Government decided to expand the Shoolpaneshwar Wildlife Sanctuary near the dam, south of the river. There is a harebrained scheme that envisages drowning animals from the submerged forests swimming their way to 'wildlife corridors' that will be created for them, and setting up home in the New! Improved! Shoolpaneshwar Sanctuary.

Presumably wildlife and biodiversity can be protected and maintained only if human activity is restricted and traditional rights to use forest resources curtailed. Forty thousand Adivasis from 101 villages within the boundaries of the Shoolpaneshwar Sanctuary depend on the forest for a livelihood. They will be 'persuaded' to leave.

They are not included in the definition of 'Project-Affected'.

Where will they go? I imagine you know by now.

Whatever their troubles in the real world, in the Shoolpaneshwar Sanctuary Interpretation Centre (where an old stuffed leopard and a mouldy sloth bear have to make do with a shared corner) the Adivasis have a whole room to themselves. On the walls there are clumsy wooden carvings, Government approved Adivasi art, with signs that say 'TRIBAL ART'. In the centre, there is a life-sized thatched hut with the door open. The pot's on the fire, the dog is asleep on the floor and all's well with the world. Outside, to welcome you, are

Mr and Mrs Adivasi. A lumpy, papier-mâché couple, smiling.
Smiling. They're not even permitted the grace of rage. That's what I can't get over.

Oh, but have I got it wrong? What if they're smiling with National Pride? Brimming with the joy of having sacrificed their lives to bring drinking water to thirsty millions in Gujarat?

*

For twenty years now, the people of Gujarat have waited for the water they believe the Wonder Canal will bring them. For years the Government of Gujarat has invested 85 per cent of the State's irrigation budget into the Sardar Sarovar Projects. Every smaller, quicker, local, more feasible scheme has been set aside for the sake of this. Election after election has been contested and won on the 'water ticket'. Everyone's hopes are pinned to the Wonder Canal. Will she fulfil Gujarat's dreams?

From the Sardar Sarovar dam, the Narmada flows through 180 kilometres of rich lowland into the Arabian sea in Bharuch. What the Wonder Canal does, more or less, is to re-route most of the river, bending it almost ninety degrees northward. It's a pretty drastic thing to do to a river. The Narmada estuary in Bharuch is one of the last known breeding places of the hilsa, probably the hottest contender for India's favourite fish.

The Stanley dam wiped out Hilsa from the Cauvery River in south India, and Pakistan's Ghulam Mohammed dam destroyed its spawning area on the Indus. Hilsa, like the salmon, is an anadromous fish – born in fresh water, migrating to the ocean as a smolt and returning to the river to spawn. The drastic reduction in water flow, the change in the chemistry of the water because of all the sediment trapped *behind* the dam, will radically alter the ecology of the estuary and modify the delicate balance of freshwater and seawater which is bound to affect the spawning. At present, the Narmada estuary produces 13,000 tonnes of Hilsa and freshwater prawn (which also breeds in brackish water). Ten thousand fisher families depend on it for a living.[68]

The Morse Committee was appalled to discover that no studies had been done of the downstream environment[69] – no documentation of the riverine ecosystem, its seasonal changes, biological species or the pattern of how its resources are used. The dam-builders had no idea what the impact of the dam would be on the people and the environment downstream, let alone any ideas on what steps to take to mitigate it.

The Government simply says that it will alleviate the loss of Hilsa fisheries by stocking the reservoir with hatchery-bred fish. (Who'll control the reservoir? Who'll grant the commercial fishing to its favourite paying customers?) The only hitch is that so far, scientists have not managed to breed Hilsa artificially. The rearing of Hilsa depends on getting spawn from wild adults, which will in all likelihood be eliminated by the dam. Dams have either eliminated or endangered one-fifth of the world's freshwater fish.[70]

So! Quiz question – where will the 40,000 fisherfolk go?

E.mail your answers to the Government that Cares dot com.

At the risk of losing readers – I've been warned

several times. 'How can you write about *irrigation*? Who the *hell* is interested?' – let me tell you what the Wonder Canal is and what she's meant to achieve. *Be* interested, if you want to snatch your future back from the sweaty palms of the Iron Triangle.

Most rivers in India are monsoon-fed. Between 80 and 85 per cent of the flow takes place during the rainy months – usually between June and September. The purpose of a dam, an irrigation dam, is to store monsoon water in its reservoir and then use it judiciously for the rest of the year, distributing it across dry land through a system of canals. The area of land irrigated by the canal network is called the 'command area'.

How will the command area, accustomed only to seasonal irrigation, its entire ecology designed for that single pulse of monsoon rain, react to being irrigated the whole year round? Perennial irrigation does to soil roughly what anabolic steroids do to the human body. Steroids can turn an ordinary athlete into an Olympic medal-winner; perennial irrigation can convert soil which produced only a single crop a year, into soil that yields *several* crops a year. Lands on which farmers traditionally grew crops that don't need a great deal of water (maize, millet, barley, and a whole range of pulses) suddenly yield water-guzzling cash crops – cotton, rice, soya bean, and the biggest guzzler of all (like those finned 1950s cars), sugar-cane. This completely alters traditional crop patterns in the command area. People stop growing things that they can afford to *eat*, and start growing things that they can only afford to *sell*. By linking themselves to the 'market' they lose control over their lives.

Ecologically too this is a poisonous payoff. Even if the markets hold out, the soil doesn't. Over time it becomes too poor to support the extra demands made on it. Gradually, in the way a steroid-using athlete becomes an invalid, the soil becomes depleted and degraded, and agricultural yields begin to decrease.[71]

In India, land irrigated by well water is today almost twice as productive as land irrigated by canals.[72] Certain kinds of soil are less suitable for perennial irrigation than others. Perennial canal irrigation raises the level of the water-table. As the water moves up through the soil, it absorbs salts. Saline water is drawn to the surface by capillary action, and the land becomes waterlogged. The 'logged' water (to coin a phrase) is then breathed into the atmosphere by plants, causing an even greater concentration of salts in the soil. When the concentration of salts in the soil reaches 1 per cent, that soil becomes toxic to plant life. This is what's called salinization.

A study[73] by the Centre for Resource and Environmental Studies at the Australian National University says that one-fifth of the world's irrigated land is salt-affected.

By the mid-1980s, 25 million of the 37 million hectares under irrigation in Pakistan were estimated to be either salinized or waterlogged or both.[74] In India the estimates vary between 6 and 10 million hectares.[75] According to 'secret' government studies[76], more than 52 per cent of the Sardar Sarovar command area is prone to waterlogging and salinization.

And that's not the end of the bad news.

The 160-kilometre-long, concrete-lined Sardar Sarovar Wonder Canal and its 75,000 kilometre network of branch canals and sub-branch canals is designed to irrigate a total of two million hectares of land spread over twelve districts. The districts of Kutch and Saurashtra (the billboards of Gujarat's Thirst campaign) are at the very tail-end of this network.

The system of canals superimposes an arbitrary concrete grid on the existing pattern of natural drainage in the command area. It's a little like reorganising the pattern of reticulate veins on the surface of a leaf. When a canal cuts across the path of a natural drain, it blocks the flow of the natural, seasonal water and leads to waterlogging. The engineering solution to this is to map the pattern of natural drainage in the area and replace it with an alternate, artificial drainage system that is built in conjunction with the canals. The problem, as you can imagine, is that doing this is enormously expensive. The cost of drainage is not included as part of the Sardar Sarovar Projects. It usually isn't, in most irrigation projects.

David Hopper, the World Bank's vice-president for South Asia, has admitted[77] that The Bank does not usually include the cost of drainage in its irrigation projects in South Asia because irrigation projects *with* adequate drainage are just too expensive. *It costs five times as much to provide adequate drainage as it does to irrigate the same amount of land.* It makes the cost of a complete Project appear unviable.

The Bank's solution to the problem is to put in the irrigation system and wait – for salinity and water-logging to set in. When all the money's spent and the land is devastated and the people are in despair, who should pop by? Why, the friendly neighbour-hood banker! And what's that bulge in his pocket? Could it be a loan for a drainage project?

In Pakistan the World Bank financed the Tarbela (1977) and Mangla dam (1967) projects on the Indus. The command areas are waterlogged.[78] Now The Bank has given Pakistan a $785 million loan for a drainage project. In India, in Punjab and Haryana it's doing the same.

Irrigation without drainage is like having a system of arteries and no veins. Pretty damn pointless.

Since the World Bank stepped back from the Sardar Sarovar Projects, it's a little unclear where the money for the drainage is going to come from. This hasn't deterred the Government from going ahead with the canal work. The result is that even before the dam is ready, before the Wonder Canal has been commissioned, before a single drop of irrigation water has been delivered, waterlogging has set in. Among the worst affected areas are the resettlement colonies.

There is a difference between the planners of the Sardar Sarovar irrigation scheme and the planners of previous projects. At least they acknowledge that water-logging and salinization are *real* problems and need to be addressed.
Their solutions, however, are corny enough to send a Hoolock Gibbon to a hooting hospital.

They plan to have a series of electronic groundwater sensors placed in every 100 square kilometres of the command area. (That works out to about 1,800 ground sensors.) These will be linked to a central computer that will analyse the data and send out commands to the canal heads to stop water flowing into areas that show signs of waterlogging. A network of 'Only-irrigation', 'Only-drainage' and 'Irrigation-cum drainage' tube-wells will be sunk, and electronically synchronised by the central computer. The saline water will be pumped out, mixed with mathematically computed quantities of fresh water and then recirculated into a network of surface and sub-surface drains (for which more land will be acquired.)[79]

To achieve the irrigation efficiency that they claim they'll achieve, according to a study done by Dr Rahul Ram for Kalpavriksh[80], 82 per cent of the water that goes into the Wonder Canal network will have to be pumped out again!

They've never implemented an electronic irrigation scheme before, not even as a pilot project. It hasn't occurred to them to experiment with some already degraded land, just to see if it works. No, they'll use our money to instal it over the whole of the 2 million hectares and *then* see if it works.

What if it doesn't? If it doesn't, it won't matter to the planners. They'll still draw the same salaries. They'll still get their pensions and their gratuity and whatever else you get when you retire from a career of inflicting mayhem on a people.

How can it possibly work? It's like sending in a rocket scientist to milk a troublesome cow. How can they manage a gigantic electronic irrigation system when they can't even line the walls of the canals without having them collapse and cause untold damage to crops and people?

When they can't even prevent the Big Dam itself from breaking off in bits when it rains?

To quote from one of their own studies '*The design, the implementation and management of the integration of groundwater and surface water in the above circumstance is complex.*'[81]

Agreed. To say the least.

Their recommendation of how to deal with the complexity: '*It will only be possible to implement such a system if all groundwater and surface water supplies are managed by a single authority.*'[82]

Aha!

It's beginning to make sense now. Who will own the water?

The Single Authority.

Who will sell the water? The Single Authority.

Who will profit from the sales? The Single Authority.

The Single Authority has a scheme whereby it will sell water by the litre, not to individuals but to farmers' co-operatives (which don't exist just yet, but no doubt the Single Authority can create Co-operatives and force farmers to co-operate).

Computer water, unlike ordinary river water, is expensive. Only those who can afford it will get it. Gradually, small farmers will get edged out by big farmers, and the whole cycle of uprootment will begin all over again.

The Single Authority, because it owns the computer water, will also decide who will grow what. It says that farmers getting computer water will not be allowed to grow sugar-cane because they'll use up the share of the thirsty millions who live at the tail-end of the canal. But the Single Authority has *already* given licences to ten large sugar mills right near the head of the canal.[83] The chief promoter of one of them is Sanat Mehta, who was chairman of the Sardar Sarovar Narmada Nigam for several years. The chief promoter of another sugar mill was Chimanbhai Patel, former chief minister of Gujarat. He (along with his wife) was the most vocal, ardent proponent of the Sardar Sarovar dam. When he died, his ashes were scattered over the dam site.

In Maharashtra, thanks to a different branch of the Single Authority, the politically powerful sugar lobby that occupies one-tenth of the state's irrigated land uses *half* the state's irrigation water.

In addition to the sugar growers, the Single Authority has recently announced a scheme[84] that envisages a series of five-star hotels, golf courses and water parks that will come up along the Wonder Canal. What earthly reason could possibly justify this? The Single Authority says it's the only way to raise money to complete the project!

I really worry about those millions of good people in Kutch and Saurashtra.

Will the water *ever* reach them?

First of all, we know that there's a lot less water in the river than the Single Authority claims there is.

Second of all, in the absence of the Narmada Sagar dam, the irrigation benefits of the Sardar Sarovar drop by a further 17 to 30 per cent.

Third of all, the irrigation efficiency of the Wonder Canal (the *actual* amount of water delivered by the system) has been arbitrarily fixed at 60 per cent. The *highest* irrigation efficiency in India, taking into account system leaks and surface evaporation, is 35 per cent.[85] This means it's likely that only *half* of the Command Area will be irrigated.
Which half? The first half.

Fourth, to get to Kutch and Saurashtra, the Wonder Canal has to negotiate its way past the ten sugar mills, the golf courses, the five-star hotels, the water parks and the cash-crop-growing, politically powerful, Patel-rich districts of Baroda, Kheda, Ahmedabad, Gandhinagar and Mehsana. (Already, in complete contravention of its own directives, the

Single Authority has allotted the city of Baroda a sizeable quantity of water.[86] When Baroda gets, can Ahmedabad be left behind? The political clout of powerful urban centres in Gujarat will ensure that they secure their share.)

Fifth, even in the (100 per cent) unlikely event that water gets there, it has to be piped and distributed to those eight thousand waiting villages.

It's worth knowing that of the one billion people in the world who have no access to safe drinking water, 855 million live in rural areas.[87] This is because the cost of installing an energy–intensive network of thousands of kilometres of pipelines, aqueducts, pumps and treatment plants that would be needed to provide drinking water to scattered rural populations is prohibitive. *Nobody* builds Big Dams to provide drinking water to rural people. Nobody can *afford* to.

When the Morse Committee first arrived in Gujarat it was impressed by the Gujarat Government's commitment to taking drinking water to such distant, rural districts.[88] The members of the Committee asked to see the detailed drinking water plans. There weren't any. (There still aren't any.)

They asked if any costs had been worked out. 'A few thousand crores,' was the breezy answer.[89] A billion dollars is an expert's calculated guess. It's not included as part of the project cost. So where is the money going to come from?

Never mind. Jus' askin'.

It's interesting that the Farakka Barrage that diverts water from the Ganga to Calcutta Port has reduced the drinking water availability for 40 million people who live downstream in Bangladesh.[90]
At times there's something so precise and mathematically chilling about nationalism.
Build a dam to take water *away* from 40 million people. Build a dam to pretend to *bring* water to 40 million people.
Who are these gods that govern us? Is there no limit to their powers?

*

The last person I met in the valley was Bhaiji Bhai. He is a Tadvi Advisi from Undava, one of the first villages where the government began to acquire land for the Wonder Canal and its 75,000-kilometre network. Bhaiji Bhai lost seventeen of his nineteen acres to the Wonder Canal. It crashes through his land, 700 feet wide including its walkways and steep, sloping embankments, like a velodrome for giant bicyclists.

The canal network affects more than two hundred thousand families. People have lost wells and trees, people have had their houses separated from their farms by the canal, forcing them to walk two or three kilometres to the nearest bridge and then two or three kilometres back along the other side. Twenty-three thousand families, let's say 100,000 people, will be, like Bhaiji Bhai, seriously affected. They don't count as 'Project-affected' and are not entitled to rehabilitation.
Like his neighbours in Kevadia Colony, Bhaiji Bhai became a pauper overnight.

Bhaiji Bhai and his people, forced to smile for photographs on government calendars. Bhaiji Bhai

and his people, denied the grace of rage. Bhaiji Bhai
and his people, squashed like bugs by this country
they're supposed to call their own.

It was late evening when I arrived at his house. We
sat down on the floor and drank oversweet tea in
the dying light. As he spoke, a memory stirred in
me, a sense of déjà vu. I couldn't imagine why. I
knew I hadn't met him before. Then I realised what
it was. I didn't recognise him, but I remembered
his story. I'd seen him in an old documentary film,
shot more than ten years ago in the valley. He was
frailer now, his beard softened with age. But his
story hadn't aged. It was still young and full of
passion. It broke my heart, the patience with which
he told it. I could tell he had told it over and over
and over again, hoping, praying, that one day, one
of the strangers passing through Undava would
turn out to be Good Luck. Or God.

Bhaiji Bhai, Bhaiji Bhai, when will you get angry?
When will you stop waiting? When will you say
'That's enough!' and reach for your weapons, whatever they may be? When will you show us the whole
of your resonant, terrifying, invincible strength?

When will you break the faith? *Will* you break the faith? Or will you let it break you?

*

To slow a beast, you break its limbs. To slow a nation, you break its people. You rob them of volition. You demonstrate your absolute command over their destiny. You make it clear that ultimately it falls to you to decide who lives, who dies, who prospers, who doesn't. To exhibit your capability you show off all that you can do, and how easily you can do it. How easily you could press a button and annihilate the earth. How you can start a war, or sue for peace. How you can snatch a river away from one and gift it to another. How you can green a desert, or fell a forest and plant one somewhere else. You use caprice to fracture a people's faith in ancient things – earth, forest, water, air.

Once that's done, what do they have left? Only you. They will turn to you, because you're all they have. They will love you even while they despise you. They will trust you even though they know you well. They will vote for you even as you squeeze the very breath from their bodies. They will drink what you give them to drink. They will breathe what you give them to breathe. They will live where you dump their belongings. They have to. What else can they do? There's no higher court of redress. You are their mother and their father. You are the judge and the jury. You are the World. You are God.

Power is fortified not just by what it destroys, but also by what it creates. Not just by what it takes, but also by what it gives. And Powerlessness reaffirmed not just by the helplessness of those who have lost, but also by the gratitude of those who have (or *think* they have) gained.

This cold, contemporary cast of power is couched between the lines of noble-sounding clauses in democratic-sounding constitutions. It's wielded by the elected representatives of an ostensibly free people. Yet no monarch, no despot, no dictator in any other century in the history of human civilisation has had access to weapons like these.

Day by day, river by river, forest by forest, mountain by mountain, missile by missile, bomb by bomb – almost without our knowing it – we are being broken.

Big Dams are to a Nation's 'Development' what Nuclear Bombs are to its Military Arsenal. They're both weapons of mass destruction. They're both weapons governments use to control their own people. Both twentieth-century emblems that mark a point in time when human intelligence has outstripped its own instinct for survival. They're both malignant indications of a civilisation turning upon itself. They represent the severing of the link, not just the link – the *understanding* – between human beings and the planet they live on. They scramble the intelligence that connects eggs to hens, milk to cows, food to forests, water to rivers, air to life and the earth to human existence.

Can we unscramble it?

Maybe. Inch by inch. Bomb by bomb. Dam by dam. Maybe by fighting specific wars in specific ways. We could begin in the Narmada valley.

This July will bring the last monsoon of the twentieth century. The ragged army in the Narmada Valley has declared that it will not move when the waters of the Sardar Sarovar reservoir rise to claim its lands and homes. Whether you love the dam or hate it, whether you want it or you don't, it is in the fitness of things that you understand the price that's being paid for it. That you have the courage to watch while the dues are cleared and the books are squared.

Our dues. Our books. Not theirs.

Be there.

May 1999

REFERENCES

1. C. V. J. Sharma (ed.), 1989. *Modern Temples of India: Selected Speeches of Jawaharlal Nehru at Irrigation and Power Projects*, pp. 40–9. Central Board of Irrigation and Power.

2. Patrick McCully, 1998. *Silenced Rivers: The Ecology and Politics of Large Dams*, p. 80. Orient Longman, Hyderabad.

3. From (uncut) film footage of Bargi dam oustees, 1995. Anurag Singh and Jharana Jhaveri, Jan Madhyam, New Delhi.

4. C. V. J. Sharma (ed.), 1989. Op cit, pp. 52–6. In a speech given before the 29th Annual Meeting of the Central Board of Irrigation and Power (17 November 1958) Nehru said, 'For some time past, however, I have been beginning to think that we are suffering from what we may call "the disease of gigantism". We want to show that we can build big dams and do big things. This is a dangerous outlook developing in India ... the idea of big – having big undertakings and doing big things for the sake of showing that we can do big things – is not a good outlook at all.' And '... It is ... the small irrigation projects, the small industries and the small plants for electric power, which will change the face of the country far more than half a dozen big projects in half a dozen places.'

5. Centre for Science and Environment, 1997. *Dying Wisdom: Rise, Fall and Potential of India's Traditional Water Harvesting Systems*, p. 399. CSE, New Delhi; Madhav Gadgil, Ramachandra Guha, 1995. *Ecology and Equity*, p. 39. Penguin India, New Delhi.

6. Indian Water Resources Society, 1998. *Five Decades of Water Resources Development in India*, p. 7.

7. World Resource Institute, 1998. *World Resources 1998–9*, p. 251. OUP, Oxford.

8. McCully, 1998. Op cit, pp. 26–9. See also *The Ecologist Asia*, Vol. 6, No. 5 (Sept.–Oct. 1998), pp. 50–51 for excerpts of speech by Bruce Babbit, US Interior Secretary, in August 1998.

9. Besides McCully, 1998, op cit, see: the CSE's *State of India's Environment*, 1999, 1985 and 1982; Nicholas Hildyard and Edward Goldsmith, 1984 *The Social and Environmental Impacts of Large Dams*, Wadebridge Ecological Centre, Cornwall, UK; Satyajit Singh, 1997. *Taming the Waters: The Political Economy of Large Dams*. OUP, New Delhi; *India: Irrigation Sector Review of the World Bank* (1991); *Large Dams: Learning from the Past, Looking to the Future*, 1997 IUCN, et al.

10. Mihir Shah & Ors, 1998. *India's Drylands: Tribal Societies and Development through Environmental Regeneration*, pp. 51–103. OUP, New Delhi.

11. Ann Danaiya Usher, 1997. *Dams as Aid: A Political Anatomy of Nordic Development Thinking*. Routledge, London and New York.

12. At current prices, Rs 2,20,000 crores, at constant 1996–7 prices.
13. G O I, 1999. *Ninth Five Year Plan 1997–2002 Vol. 2*, p. 478. Planning Commission, New Delhi.
14. D. K. Mishra & R. Rangachari, 1999. *The Embankment Trap and Some Disturbing Questions*, pp. 40–8 and 62–3 respectively, Seminar 478 (June 1999); C S E, 1991. *Floods, Floodplains and Environmental Myths*.
15. Mihir Shah & Ors, 1998. Op cit, pp. 51–103.
16. Satyajit Singh, 1997. Op cit, pp. 188–90; also, G O I figures for actual displacement.
17. At a meeting in New Delhi on 21 January 1999 organized by the Union Ministry of Rural Areas and Employment, for discussions on the draft National Resettlement and Rehabilitation Policy and the Amendment to the draft Land Acquisition Act.
18. Bradford Morse & Thomas Berger, 1992. *Sardar Sarovar; The Report of the Independent Review*, p. 62. Originally published by Resource Futures International (R F I) Inc., Ottawa.
19. G O I, 28th and 29th *Report of the Commissioner for Scheduled Castes and Scheduled Tribes*, New Delhi, 1988–9.
20. 10 April 1999 (front page), *Indian Express*, New Delhi.
21. G O I, 1999, *Ninth Five-Year Plan 1991–2002*, Vol. 2, p. 437.
22. Siddharth Dube, 1998. *Words Like Freedom*,

HarperCollins (India), New Delhi; CMIE (Centre for Monitoring the Indian Economy), 1996. See also *World Bank Poverty Update*, quoted in *Business Line*, 4 June 1999.

23. National Human Rights Commission, *Report of the Visit of the Official Team of the NHRC to the Scarcity-affected Areas of Orissa*, December 1996.

24. GOI, *Award of the Narmada Water Disputes Tribunal* 1978–9.

25. GOI, *Report of the FMG-2 on SSP* 1995; cf. various affidavits of the Government of India and Government of Madhya Pradesh before the Supreme Court of India, 1991–8.

26. CWC, *Monthly Observed Flows of the Narmada at Garudeshwar*, 1992, Hydrology Studies Organization, Central Water Commission, New Delhi.

27. *Written Submission on Behalf of Union of India*, February 1999, page 7, clause 1.7.

28. *Tigerlink News*, Vol. 5 No. 2, June 1999, p. 28.

29. *World Bank Annual Reports 1993–8*.

30. McCully, 1998. Op cit, p. 274.

31. McCully, 1998. Op cit, p. 21. The World Bank started funding dams in China in 1984. Since then, it has lent around $3.4 billion (not adjusted for inflation) to finance thirteen Big Dams that will cause the displacement of 360,000 people. The centrepiece of the World Bank's dam financing in China is the Xiaolangdi dam on the Yellow River, which will single-handedly displace 181,000 people.

32. McCully, 1998. Op cit, p. 278.

33. J. Vidal & N. Cumming-Bruce, 'The Curse of Pergau', *The Economist*, 5 March 1994; 'Dam Price Jumped 81 Million Pounds Days After Deal', *Guardian*, London, 19 January 1994; 'Whitehall Must Not Escape Scot Free', *Guardian*, London, 12 February 1994; quoted in McCully, 1968, op cit, p. 291.

34. McCully, 1998. Op cit, p. 62.

35. For example, see Sardar Sarovar Narmada Nigam Ltd, 1989. *Planning for Prosperity*; Babubhai J. Patel, 1992. *Progressing amidst Challenges*, C. C. Patel, 1991. *SSP, What It is and What It is not*; P. A. Raj, 1989, 1990 and 1991 editions. *Facts: Sardar Sarovar Project*.

36. Ibid; also Rahul Ram, 1993. *Muddy Waters: A Critical Assessment of the Benefits of the Sardar Sarovar Project*, Kalpavriksh, New Delhi.

37. Morse, 1992. Op cit, p. 319. According to official statistics (Narmada Control Authority, 1992. *Benefits to Saurashtra and Kutch Areas in Gujarat*, NCA, Indore), 948 villages in Kutch and 4877 villages in Saurashtra are to get drinking water from the SSP. However, according to the 1981 census, there are only 887 inhabited villages in Kutch and 4727 villages in the whole of Saurashtra. The planners had simply hoovered up the names of villages from a map, thereby including the names of 211 deserted villages! Cited in Rahul Ram, 1993, op cit.

38. For example, the minutes of the various meetings

of the Rehabilitation and Resettlement Sub Groups of the Narmada Control Authority, 1998–9. Also, Morse, 1992. Op cit, p. 51.

39. Rahul Ram, 1993. Op cit, p. 34.

40. See, for example, the petition filed by the NBA in the Supreme Court, 1994.

41. SSNNL, 1989. *Planning for Prosperity*, Government of Gujarat.

42. S. Dharmadhikary, 1995. *Hydropower at Sardar Sarovar: Is It Necessary Justified and Affordable?*, p. 141. In W. F. Fisher (ed.), *Towards Sustainable Development? Struggling Over India's Narmada River*, M. F. Sharpe, Armonk, New York.

43. McCully, 1998. Op cit, p. 87.

44. McCully, 1998. Op cit, p. 185.

45. World Bank, 1994. *Resettlement and Development: The Bankwide Review of Projects Involving Resettlement 1986–1993.*

46. World Bank, 1994 (ii). *Resettlement and Rehabilitation of India: A Status Update of Projects Involving Involuntary Resettlement.*

47. World Bank, 1994. *Resettlement and Development*, op cit.

48. Morse, 1992. Op cit, Letter to the President, pp. XII, XXIV and XXV.

49. Morse, 1992. Op cit, p. XXV.

50. Minimum conditions included unfinished appraisal of social and environmental impacts. For details, see Udall, *The International Narmada Campaign*;

McCully, 1992; *Cracks in the Dam: The World Bank in India*, Multinational Monitor, December 1992.

51. See the letter from the GOI to the World Bank, 29 March 1993; press release of the World Bank dated 30 March 1993, a copy of which can be found in Campaign Information Package of International Rivers Network, *Narmada Valley Development Project*, Vol. 1, August 1998.

52. The date was 14 November 1992. Venue: outside the Taj Mahal Hotel, Bombay, where Lewis Preston, President of the World Bank, was staying. See Lawyers Committee for Human Rights, April 1993. *Unacceptable Means: India's Sardar Sarovar Project and Violations of Human Rights: Oct. 1992–Feb. 1993*, pp. 10–12.

53. On the night of 20 March 1994, the NBA Office at Baroda was attacked by hoodlums simply because of a (baseless) rumour that one member of the Five Member Group Committee was sitting inside with members of the NBA. Some NBA activists were manhandled, and a large collection of NBA documents was burnt and destroyed.

54. Ministry of Water Resources, GOI, 1994. *Report of the Five Member Group on Sardar Sarovar Project*.

55. Writ Petition 319 of 1994 argued that the Sardar Sarovar Project violated the fundamental rights of those affected by the project, and that the project was not viable on social, environmental, technical

(including seismic and hydrological), financial or economic grounds. The Writ Petition asked for a comprehensive review of the project, pending which construction on the project should cease.

56. *Frontline*, 27 January 1995; *Sunday*, 21 January 1995.

57. In January 1995, the Supreme Court took on record the statement of the Counsel for the Union of India that no further work on the Sardar Sarovar dam would be done without informing the Court in advance. On 4 May 1995, the Court allowed construction of 'humps' on the dam, on the plea of the Union of India that they were required for reasons of safety. The Court, however, reiterated its order of January 1995 that no further construction will be done without the express permission of the Court.

58. *Report of the Narmada Water Disputes Tribunal with Its Decision*, Vol. II 1979, p. 102; cited in Morse, 1992. Op cit, p. 250.

59. Morse, 1992, Op cit, pp. 323–9.

60. P. A. Raj, 1989, 1990, 1991, *Facts: Sardar Sarovar Project*, Sardar Sarovar Narmada Nigam Ltd, Gujarat.

61. Medha Patkar, 1995. 'The Struggle for Participation and Justice: A Historical Narrative', in Fisher William (ed.), *Toward Sustainable Development: Struggling over India's Narmada River*, M. Sharpe, Inc., pp. 159–78; S. Parasuraman, 1997.

'The Anti-Dam Movement and Rehabilitation Policy', in Jean Dreze et al., *The Dam and the Nation*, OUP, pp. 26–65; minutes of various meetings of the R & R sub-group of the Narmada Control Authority.

62. On my visit to the valley in March 1999, I was told this by villagers at Mokhdi who had returned from their resettlement colonies.

63. *Kaise Jeebo Re*, documentary film by Anurag Singh and Jharana Jhaveri, Jan Madhyam 1997; also, unedited footage in the NBA archives.

64. Letter to the Independent Review from a resident of Paryeta Resettlement Colony, cited in Morse, 1992. Op cit, pp. 159–60.

65. *Narmada Manavadhikar Yatra*, that travelled from the Narmada valley to Delhi via Bombay. It reached Delhi on 7 April 1999.

66. Told to me by Mohan Bhai Tadvi, in Kevadia Colony, March 1999.

67. Morse, 1992. Op cit, pp. 89–94; NBA interviews, March 1999.

68. NBA interviews, March 1999.

69. Morse, 1992. Op cit, pp. 277–94.

70. McCully, 1998. Op cit, pp. 46–9.

71. For a discussion on the subject, see the World Bank, 1991, *India Irrigation Sector Review*; A. Vaidyanathan, 1994. *Food, Agriculture and Water*, MIDS, Madras; McCully, 1998. Op cit, pp. 182–207.

72. The World Bank, 1991. *India Irrigation Sector Review*, Vol. 2, p. 7.

73. Cited in McCully, 1998. Op cit, p. 187.

74. Shaheen Rafi Khan, 1998. *The Kalabagh Controversy*, Sustainable Development Policy Institute, Pakistan; E. Goldsmith 1998. 'Learning to Live with Nature: The Lessons of Traditional Irrigation' in *The Ecologist*, Vol. 6, No. 5, Sept./Oct. 1998.

75. Mihir Shah & Ors, 1998. Op cit, p. 51; also in Goldsmith, 1998. Op cit.

76. Operations Research Group, 1981. *Critical Zones in Narmada Command – Problems and Prospects*, ORG, Baroda; ORG, 1982. *Regionalisation of Narmada Command*, ORG, Gandhinagar; World Bank, 1985. *Staff Appraisal Report, India, Narmada River Development – Gujarat, Water Delivery and Drainage Project*, Report No. 5108-IN; Core Consultants, 1982. *Main Report: Narmada Mahi Doab Drainage Study,* commissioned by Narmada Planning Group, Government of Gujarat.

77. Robert Wade, 1997. 'Greening the Bank: The Struggle over the Environment, 1970–1995', pp. 661–2 in Devesh Kapur et al (eds), *The World Bank: Its First Half Century*, Brookings Institution Press, Washington DC.

78. Shaheen Rafi Khan, 1998, op cit.

79. CES, 1992. *Pre-Feasibility Level Drainage Study for SSP Command Beyond River Mahi*, CES Water Resources Development and Management

Consultancy Pvt Ltd, New Delhi, for Government of Gujarat.

80. Rahul Ram, 1995. 'The Best-laid Plans . . .', p. 78 in *Frontline*, 14 July 1995.

81. Core Consultants, 1982. Op cit, p. 66.

82. Ibid.

83. For example, see GOI, 1995, *Report of the FMG*; or Rahul Ram, 1993, op cit.

84. Called the 'Economic Regeneration Programme', formulated to generate funds for the cash-strapped Sardar Sarovar Narmada Nigam Ltd. Under the programme, land along the main canal of the Narmada Project will be acquired and sold for tourist facilities, hotels, water parks, fun world sites, garden restaurants, etc. Cf. *The Times of India* (Ahmedabad), 17 May 1998.

85. World Bank, 1991 *India Irrigation Sector Review*.

86. Written submissions on behalf of the petitioners (NBA) in the Supreme Court, January 1999, p. 63; *The Times of India* (Ahmedabad), 23 May 1999.

87. Ismail Serageldin, 1994. *Water Supply, Sanitation and Environmental Sustainability*, p. 4. The World Bank, Washington DC.

88. Morse, 1992. Op cit, p. xxiii.

89. Morse, 1992. Op cit, pp. 317–19.

90. McCully, 1998. Op cit, p. 167.

The End of Imagination

For
marmots and voles
and everything else on the earth
that is threatened and terrorized
by the human race.

'The desert shook,' the Government of India informed us (its people).
'The whole mountain turned white,' the Government of Pakistan replied.

'By afternoon the wind had fallen silent over Pokhran. At 3.45 p.m. the timer detonated the three devices. Around 200 to 300 m. deep in the earth, the heat generated was equivalent to a million degrees centigrade – as hot as temperatures on the sun. Instantly, rocks weighing around a thousand tons, a mini mountain underground, vaporized . . . shockwaves from the blast began to lift a mound of earth the size of a football field by several metres. One scientist on seeing it said, "I can now believe stories of Lord Krishna lifting a hill".'

India Today

May 1998. It'll go down in history books, provided of course we have history books to go down in. Provided, of course, we have a future. There's nothing new or original left to be said about nuclear weapons. There can be nothing more humiliating for a writer of fiction to have to do than restate a case that has, over the years, already been made by other people in other parts of the world, and made passionately, eloquently and knowledgeably.

I am prepared to grovel. To humiliate myself abjectly, because, in the circumstances, silence would be indefensible. So those of you who are willing: let's pick our parts, put on these discarded costumes and speak our second-hand lines in this sad second-hand play. But let's not forget that the stakes we're playing for are huge. Our fatigue and our shame could mean the end of us. The end of our children and our children's children. Of everything we love. We have to reach within ourselves and find the strength to think. To fight.

Once again we are pitifully behind the times – not just scientifically and technologically (ignore the hollow claims), but more pertinently in our ability to grasp the true nature of nuclear weapons. Our Comprehension of the Horror Department is hopelessly obsolete. Here we are, all of us in India and in Pakistan, discussing the finer points of politics and foreign policy, behaving for all the world as though our governments have just devised a newer, bigger bomb, a sort of immense hand grenade with which they will annihilate the enemy (each other) and protect us from all harm. How desperately we want to believe that. What wonderful, willing, well-behaved, gullible subjects we have turned out to be. The rest of humanity (Yes, yes, I know, I *know*, but let's ignore Them for the moment. They forfeited their votes a long time ago), the rest of the rest of humanity may not forgive us, but then the rest of the rest of humanity, depending on who fashions its views, may not know what a tired, dejected heart-broken people we are. Perhaps it doesn't realize how urgently we need a miracle. How deeply we yearn for magic.

If only, if *only*, nuclear war was just another kind of war. If only it was about the usual things – nations and territories, gods and histories. If only those of us who dread it are just worthless moral cowards who are not prepared to die in defence of our beliefs. If only nuclear war was the kind of war in which countries battle countries and men battle men. But it isn't. If there is a nuclear war, our foes will not be China or America or even each other.

Our foe will be the earth herself. The very elements – the sky, the air, the land, the wind and water – will all turn against us. Their wrath will be terrible.

Our cities and forests, our fields and villages will burn for days. Rivers will turn to poison. The air will become fire. The wind will spread the flames. When everything there is to burn has burned and the fires die, smoke will rise and shut out the sun. The earth will be enveloped in darkness. There will be no day. Only interminable night. Temperatures will drop to far below freezing and nuclear winter will set in. Water will turn into toxic ice. Radioactive fallout will seep through the earth and contaminate groundwater. Most living things, animal and vegetable, fish and fowl, will die. Only rats and cockroaches will breed and multiply and compete with foraging, relict humans for what little food there is.

What shall we do then, those of us who are still alive? Burned and blind and bald and ill, carrying the cancerous carcasses of our children in our arms, where shall we go? What shall we eat? What shall we drink? What shall we breathe?

The Head of the Health, Environment and Safety Group of the Bhabha Atomic Research Centre in Bombay has a plan. He declared in an interview (*The Pioneer*, 24 April 1998) that India could survive nuclear war. His advice is that if there is a nuclear war, we take the same safety measures as the ones that scientists have recommended in the event of accidents at nuclear plants.

Take iodine pills, he suggests. And other steps such as remaining indoors, consuming only stored water and food and avoiding milk. Infants should be given powdered milk. 'People in the danger zone should immediately go to the ground floor and if possible to the basement.'

What do you do with these levels of lunacy? What do you do if you're trapped in an asylum and the doctors are all dangerously deranged?

Ignore it, it's just a novelist's naïveté, they'll tell you, Doomsday Prophet hyperbole. It'll never come to that. There will *be* no war. Nuclear weapons are about peace, not war. 'Deterrence' is the buzzword of the people who like to think of themselves as hawks. (Nice birds, those. Cool. Stylish. Predatory. Pity there won't be many of them around after the war. Extinction is a word we must try and get used to.) Deterrence is an old thesis that has been resurrected and is being recycled with added local flavour. The Theory of Deterrence cornered the credit for having prevented the Cold War from turning into a Third World War. The only immutable fact about the Third World War is that if there's going to be one, it will be fought after the Second World War. In other words, there's no fixed schedule. In other words, we still have time. And perhaps the pun (the Third World War) is prescient. True, the Cold War is over, but let's not be hoodwinked by the ten-year lull in nuclear posturing. It was just a cruel joke. It was only in remission. It wasn't cured. It proves no theories. After all, what is ten years in the history of the world? Here it is again, the disease. More widespread and less amenable to any sort of treatment than ever. No, the Theory of Deterrence has some fundamental flaws.

*

Flaw Number One is that it presumes a complete, sophisticated understanding of the psychology of your enemy. It assumes that what deters you (the fear of annihilation) will deter them. What about those who are *not* deterred by that? The suicide-bomber psyche – the 'We'll take you with us' school – is that an outlandish thought? How did Rajiv Gandhi die?

In any case who's the 'you' and who's the 'enemy'? Both are only governments. Governments change. They wear masks within masks. They moult and reinvent themselves all the time. The one we have at the moment, for instance, does not even have enough seats to last a full term in office, but demands that we trust it to do pirouettes and party tricks with nuclear bombs even as it scrabbles around for a foothold to maintain a simple majority in Parliament.

Flaw Number Two is that Deterrence is premised on fear. But fear is premised on knowledge. On an understanding of the true extent and scale of the devastation that nuclear war will wreak. It is not some inherent, mystical attribute of nuclear bombs that they automatically inspire thoughts of peace. On the contrary, it is the endless, tireless, confron-

tational work of people who have had the courage
openly to denounce them, the marches, the demon-
strations, the films, the outrage – *that* is what has
averted, or perhaps only postponed, nuclear war.
Deterrence will not and cannot work given the
levels of ignorance and illiteracy that hang over
our two countries like dense, impenetrable veils.
(Witness the VHP wanting to distribute radioactive sand from the Pokhran desert as prasad all
across India. A cancer yatra?) The Theory of
Deterrence is nothing but a perilous joke in a world
where iodine pills are prescribed as a prophylactic
for nuclear irradiation.

India and Pakistan have nuclear bombs now and
feel entirely justified in having them. Soon others
will too. Israel, Iran, Iraq, Saudi Arabia, Norway,
Nepal (I'm trying to be eclectic here), Denmark,
Germany, Bhutan, Mexico, Lebanon, Sri Lanka,
Burma, Bosnia, Singapore, North Korea, Sweden,
South Korea, Vietnam, Cuba, Afghanistan, Uzbek-
istan . . . and why not? Every country in the world
has a special case to make. Everybody has borders
and beliefs. And when all our larders are bursting
with shiny bombs and our bellies are empty (deterrence is an exorbitant beast), we can trade bombs

for food. And when nuclear technology goes on the market, when it gets truly competitive and prices fall, not just governments, but anybody who can afford it can have their own private arsenal – businessmen, terrorists, perhaps even the occasional rich writer (like myself). Our planet will bristle with beautiful missiles. There will be a new world order. The dictatorship of the pro-nuke elite. We can get our kicks by threatening each other. It'll be like bungee-jumping when you can't rely on the bungee cord, or playing Russian roulette all day long. An additional perk will be the thrill of Not Knowing What to Believe. We can be victims of the predatory imagination of every green card-seeking charlatan who surfaces in the West with concocted stories of imminent missile attacks. We can delight at the prospect of being held to ransom by every petty trouble-maker and rumour-monger, the more the merrier if truth be told, anything for an excuse to make more bombs. So you see, even without a war, we have a lot to look forward to.

But let us pause to give credit where it's due. Whom must we thank for all this?

The Men who made it happen. The Masters of the Universe. Ladies and gentlemen, the United States of America! Come on up here folks, stand up and take a bow. Thank you for doing this to the world. Thank you for making a difference. Thank you for showing us the way. Thank you for altering the very meaning of life.

From now on it is not dying we must fear, but living.

It is such supreme folly to believe that nuclear weapons are deadly only if they're used. The fact that they exist at all, their very presence in our lives, will wreak more havoc than we can begin to fathom. Nuclear weapons pervade our thinking. Control our behaviour. Administer our societies. Inform our dreams. They bury themselves like meat hooks deep in the base of our brains. They are purveyors of madness. They are the ultimate colonizer. Whiter than any white man that ever lived. The very heart of whiteness.

All I can say to every man, woman and sentient child here in India, and over there, just a little way away in Pakistan, is: take it personally. Whoever you are – Hindu, Muslim, urban, agrarian – it doesn't matter. The only good thing about nuclear war is that it is the single most egalitarian idea that man has ever had. On the day of reckoning, you will not be asked to present your credentials. The devastation will be undiscriminating. The bomb isn't in your back yard. It's in your body. And mine. *Nobody*, no nation, no government, no man, no god, has the right to put it there. We're radioactive already, and the war hasn't even begun. So stand up and say something. Never mind if it's been said before. Speak up on your own behalf. Take it very personally.

THE BOMB AND I

In early May (before the bomb), I left home for three weeks. I thought I would return. I had every intention of returning. Of course, things haven't worked out quite the way I had planned.

While I was away, I met a friend of mine whom I have always loved for, among other things, her ability to combine deep affection with a frankness that borders on savagery.

'I've been thinking about you,' she said, 'about *The God of Small Things* – what's in it, what's over it, under it, around it, above it . . .'

She fell silent for a while. I was uneasy and not at all sure that I wanted to hear the rest of what she had to say. She, however, was sure that she was going to say it. 'In this last year – less than a year actually – you've had too much of everything – fame, money, prizes, adulation, criticism, condemnation, ridicule, love, hate, anger, envy, generosity – everything. In some ways it's a perfect story. Perfectly baroque in its excess. The trouble is that it has, or can have, only one perfect ending.' Her eyes were on me, bright with a slanting, probing brilliance. She knew that I knew what she was going to say. She was insane.

She was going to say that nothing that happened to me in the future could ever match the buzz of this. That the whole of the rest of my life was going to be vaguely unsatisfying. And, therefore, the only perfect ending to the story would be death. *My* death.

The thought had occurred to me too. Of course it had. The fact that all this, this global dazzle – these lights in my eyes, the applause, the flowers, the photographers, the journalists feigning a deep interest in my life (yet struggling to get a single fact straight), the men in suits fawning over me, the shiny hotel bathrooms with endless towels – none of it was likely to happen again. Would I miss it? Had I grown to need it? Was I a fame-junkie? Would I have withdrawal symptoms?

The more I thought about it, the clearer it became to me that if fame was going to be my permanent condition it would kill me. Club me to death with its good manners and hygiene. I'll admit that I've enjoyed my own five minutes of it immensely, but primarily *because* it was just five minutes. Because I knew (or thought I knew) that I could go home when I was bored and giggle about it. Grow old and irresponsible. Eat mangoes in the moonlight. Maybe write a couple of failed books – worstsellers – to see what it felt like. For a whole year I've cartwheeled across the world, anchored always to thoughts of home and the life I would go back to. Contrary to all the enquiries and predictions about my impending emigration, that was the well I dipped into. That was my sustenance. My strength. I told my friend there was no such thing as a perfect story. I said in any case hers was an external view of things, this assumption that the trajectory of a person's happiness, or let's say fulfilment, had peaked (and now must trough) because she had accidentally stumbled upon 'success'. It was premised on the unimaginative belief that wealth and fame were the mandatory stuff of everybody's dreams.

You've lived too long in New York, I told her. There are other worlds. Other kinds of dreams. Dreams in which failure is feasible. Honourable. Sometimes even worth striving for. Worlds in which recognition is not the only barometer of brilliance or human worth. There are plenty of warriors that I know and love, people far more valuable than myself, who go to war each day, knowing in advance that they will fail. True, they are less 'successful' in the most vulgar sense of the word, but by no means less fulfilled.

The only dream worth having, I told her, is to dream that you will live while you're alive and die only when you're dead. (Prescience? Perhaps.)

'Which means exactly what?' (Arched eyebrows, a little annoyed.)

I tried to explain, but didn't do a very good job of it. Sometimes I need to write to think. So I wrote it down for her on a paper napkin. This is what I wrote: *To love. To be loved. To never forget your own insignificance. To never get used to the unspeakable violence and the vulgar disparity of life around you. To seek joy in the saddest places. To pursue beauty to its lair.*

To never simplify what is complicated or complicate what is simple. To respect strength, never power. Above all, to watch. To try and understand. To never look away. And never, never to forget.

I've known her for many years, this friend of mine. She's an architect too.

She looked dubious, somewhat unconvinced by my paper napkin speech. I could tell that structurally, just in terms of the sleek, narrative symmetry of things, and because she loved me, her thrill at my 'success' was so keen, so generous, that it weighed in evenly with her (anticipated) horror at the idea of my death. I understood that it was nothing personal. Just a design thing.

*

Anyhow, two weeks after that conversation, I returned to India. To what I think/thought of as home. Something had died but it wasn't me. It was infinitely more precious. It was a world that has been ailing for a while, and has finally breathed its last. It's been cremated now. The air is thick with ugliness and there's the unmistakable stench of fascism on the breeze.

Day after day, in newspaper editorials, on the radio, on TV chat shows, on MTV for heaven's sake, people whose instincts one thought one could trust – writers, painters, journalists – make the crossing. The chill seeps into my bones as it becomes painfully apparent from the lessons of everyday life that what you read in history books is true. That fascism is indeed as much about people as about governments. That it begins at home. In drawing rooms. In bedrooms. In beds. 'Explosion of self-esteem', 'Road to Resurgence', 'A Moment of Pride', these were headlines in the papers in the days following the nuclear tests. 'We have proved that we are not eunuchs any more,' said Mr Thackeray of the Shiv Sena. (Whoever said we were? True, a good number of us are women, but that, as far as I know, isn't the same thing.) Reading the papers, it was often hard to tell when people were referring to Viagra (which was competing for second place on the front pages) and when they were talking about the bomb – 'We have superior strength and potency.' (This was our Minister for Defence after Pakistan completed its tests.)

'These are not just nuclear tests, they are nationalism tests,' we were repeatedly told.

This has been hammered home, over and over again. The bomb is India. India is the bomb. Not just India, Hindu India. Therefore, be warned, any criticism of it is not just anti-national, but anti-Hindu. (Of course, in Pakistan the bomb is Islamic. Other than that, politically, the same physics applies.) This is one of the unexpected perks of having a nuclear bomb. Not only can the Government use it to threaten the Enemy, they can use it to declare war on their own people. Us.

In 1975, one year after India first dipped her toe into the nuclear sea, Mrs Gandhi declared the Emergency. What will 1999 bring? There's talk of cells being set up to monitor anti-national activity. Talk of amending cable laws to ban networks 'harming national culture' (*The Indian Express*, 3 July). Of churches being struck off the list of religious places because 'wine is served' (announced and retracted, *The Indian Express*, 3 July, *The Times of India*, 4 July). Artists, writers, actors and singers are being harassed, threatened (and are succumbing to the threats). Not just by goon squads, but by instruments of the government. And in courts of law. There are letters and articles circulating on the Net – creative interpretations of Nostradamus' predictions claiming that a mighty, all-conquering Hindu nation is about to emerge – a resurgent India that will 'burst forth upon its former oppressors and destroy them completely'. That 'the beginning of the terrible revenge (that will wipe out all Moslems) will be in the seventh month of 1999'. This may well be the work of some lone nut, or a bunch of arcane god-squadders. The trouble is that having a nuclear bomb makes thoughts like these seem feasible. It *creates* thoughts like these. It bestows on people these utterly misplaced, utterly deadly notions of their own power. It's happening. It's all happening. I wish I could say 'slowly but surely' – but I can't. Things are moving at a pretty fair clip.

Why does it all seem so familiar? Is it because, even as you watch, reality dissolves and seamlessly rushes forward into the silent, black-and-white images from old films – scenes of people being hounded out of their lives, rounded up and herded into camps? Of massacre, of mayhem, of endless columns of broken people making their way to nowhere? Why is there no soundtrack? Why is the hall so quiet? Have I been seeing too many films? Am I mad? Or am I right? Could those images be the inevitable culmination of what we have set into motion? Could our future be rushing forward into our past? I think so. Unless, of course, nuclear war settles it once and for all.

When I told my friends that I was writing this piece, they cautioned me. 'Go ahead,' they said, 'but first make sure you're not vulnerable. Make sure your papers are in order. Make sure your taxes are paid.'

My papers are in order. My taxes are paid. But how can one *not* be vulnerable in a climate like this? Everyone is vulnerable. Accidents happen. There's safety only in acquiescence. As I write, I am filled with foreboding. In this country, I have truly known what it means for a writer to feel loved (and, to some degree, hated too). Last year I was one of the items being paraded in the media's end-of-the-year National Pride Parade. Among the others, much to my mortification, were a bomb-maker and an international beauty queen. Each time a beaming person stopped me on the street and said 'You have made India proud' (referring to the prize I won, not the book I wrote), I felt a little uneasy. It frightened me then and it terrifies me now, because

I know how easily that swell, that tide of emotion, can turn against me. Perhaps the time for that has come. I'm going to step out from under the fairy lights and say what's on my mind.

It's this:

If protesting against having a nuclear bomb implanted in my brain is anti-Hindu and anti-national, then I secede. I hereby declare myself an independent, mobile republic. I am a citizen of the earth. I own no territory. I have no flag. I'm female, but have nothing against eunuchs. My policies are simple. I'm willing to sign any nuclear non-proliferation treaty or nuclear test ban treaty that's going. Immigrants are welcome. You can help me design our flag.

My world has died. And I write to mourn its passing.

Admittedly it was a flawed world. An unviable world. A scarred and wounded world. It was a world that I myself have criticized unsparingly, but only because I loved it. It didn't deserve to die. It didn't deserve to be dismembered. Forgive me, I realize that sentimentality is uncool – but what shall I do with my desolation?

I loved it simply because it offered humanity a choice. It was a rock out at sea. It was a stubborn chink of light that insisted that there was a different way of living. It was a functioning possibility. A real option. All that's gone now. India's nuclear tests, the manner in which they were conducted, the euphoria with which they have been greeted (by us) is indefensible. To me, it signifies dreadful things. The end of imagination. The end of freedom actually, because, after all, that's what freedom is. Choice.

On 15 August last year we celebrated the fiftieth anniversary of India's independence. In May we can mark our first anniversary in nuclear bondage.

Why did they do it?

Political expediency is the obvious, cynical answer, except that it only raises another, more basic question: why should it have been politically expedient?

The three Official Reasons given are: China, Pakistan and Exposing Western Hypocrisy.

Taken at face value, and examined individually, they're somewhat baffling. I'm not for a moment suggesting that these are not real issues. Merely that they aren't new. The only new thing on the old horizon is the Indian Government. In his appallingly cavalier letter to the US President (why bother to write at all if you're going to write like this?) our Prime Minister says India's decision to go ahead with the nuclear tests was due to a 'deteriorating security environment'. He goes on to mention the war with China in 1962 and the 'three aggressions we have suffered in the last fifty years (from Pakistan). And for the last ten years we have been the victim of unremitting terrorism and militancy sponsored by it ... especially in Jammu and Kashmir.'

The war with China is thirty-five years old. Unless there's some vital state secret that we don't know about, it certainly seemed as though matters had improved slightly between us. Just a few days before the nuclear tests General Fu Quanyou, Chief of General Staff of the Chinese People's Liberation Army, was the guest of our Chief of Army Staff. We heard no words of war.

The most recent war with Pakistan was fought twenty-seven years ago. Admittedly Kashmir continues to be a deeply troubled region and no doubt Pakistan is gleefully fanning the flames. But surely there must be flames to fan in the first place? Surely the kindling is crackling and ready to burn? Can the Indian State with even a modicum of honesty absolve itself completely of having a hand in Kashmir's troubles? Kashmir, and for that matter, Assam Tripura, Nagaland – virtually the whole of the northeast – Jharkhand, Uttarakhand and all the trouble that's still to come – these are symptoms of a deeper malaise. It cannot and will not be solved by pointing nuclear missiles at Pakistan.

Even Pakistan can't be solved by pointing nuclear missiles at Pakistan. Though we are separate countries, we share skies, we share winds, we share water. Where radioactive fallout will land on any given day depends on the direction of the wind and rain. Lahore and Amritsar are thirty miles apart. If we bomb Lahore, Punjab will burn. If we bomb Karachi, then Gujarat and Rajasthan, perhaps even Bombay, will burn. Any nuclear war with Pakistan will be a war against ourselves.

As for the third Official Reason: Exposing Western Hypocrisy – how much more exposed can they be? Which decent human being on earth harbours any illusions about it? These are people whose histories are spongy with the blood of others. Colonialism, apartheid, slavery, ethnic cleansing, germ warfare, chemical weapons – they virtually invented it all. They have plundered nations, snuffed out civilizations, exterminated entire populations. They stand on the world's stage stark naked but entirely unembarrassed, because they know that they have more money, more food and bigger bombs than anybody else. They know they can wipe us out in the course of an ordinary working day. Personally, I'd say it is more arrogance than hypocrisy.

*

We have less money, less food and smaller bombs. However, we have, or had, all kinds of other wealth. Delightful, unquantifiable. What we've done with it is the opposite of what we think we've done. We've pawned it all. We've traded it in. For what? In order to enter into a contract with the very people we claim to despise. In the larger scheme of things, we've agreed to play their game and play it their way. We've accepted their terms and conditions unquestioningly. The CTBT ain't nothin' compared to this.

All in all, I think it is fair to say that *we're* the hypocrites. We're the ones who've abandoned what was arguably a moral position, i.e.: *We have the technology, we can make bombs if we want to, but we won't. We don't believe in them.*

We're the ones who have now set up this craven clamouring to be admitted into the club of Superpowers. (If we are, we will no doubt gladly slam the door after us, and say to hell with principles about fighting Discriminatory World Orders.) For India to demand the status of a Superpower is as ridiculous as demanding to play in the World Cup finals simply because we have a ball. Never mind that we haven't qualified, or that we don't play much soccer and haven't got a team.

Since we've chosen to enter the arena, it might be an idea to begin by learning the rules of the game. Rule number one is Acknowledge the Masters. Who are the best players? The ones with more money, more food, more bombs.

Rule number two is Locate Yourself in Relation to Them, i.e.: make an honest assessment of your position and abilities. The honest assessment of ourselves (in quantifiable terms) reads as follows:

We are a nation of nearly a billion people. In development terms we rank No. 138 out of the 175 countries listed in the UNDP's Human Development Index. More than 400 million of our people are illiterate and live in absolute poverty, over 600 million lack even basic sanitation and over 200 million have no safe drinking water.

So the three Official Reasons, taken individually, don't hold much water. However, if you link them, a kind of twisted logic reveals itself. It has more to do with us than them.

The key words in our Prime Minister's letter to the US President were 'suffered' and 'victim'. That's the substance of it. That's our meat and drink. We *need* to feel like victims. We need to feel beleaguered. We need enemies. We have so little sense of ourselves as a nation and therefore constantly cast about for targets to define ourselves against. Prevalent political wisdom suggests that to prevent the State from crumbling, we need a national cause, and other than our currency (and, of course, poverty, illiteracy and elections), we have none. This is the heart of the matter. This is the road that has led us to the bomb. This search for selfhood. If we are looking for a way out, we need some honest answers to some uncomfortable questions. Once again, it isn't as though these questions haven't been asked before. It's just that we prefer to mumble the answers and hope that no one's heard.

Is there such a thing as an Indian identity?

Do we really need one?

Who is an authentic Indian and who isn't?
Is India Indian?
Does it matter?

Whether or not there has ever been a single civilization that could call itself 'Indian Civilization',
whether or not India was, is, or ever will become
a cohesive cultural entity, depends on whether you
dwell on the differences or the similarities in the
cultures of the people who have inhabited the subcontinent for centuries. India, as a modern nation
state, was marked out with precise geographical
boundaries, in their precise geographical way, by a
British Act of Parliament in 1899. Our country, as
we know it, was forged on the anvil of the British
Empire for the entirely unsentimental reasons of
commerce and administration. But even as she was
born, she began her struggle against her creators.
So is India Indian? It's a tough question. Let's just
say that we're an ancient people learning to live in
a recent nation.

What is true is that India is an artificial State – a
State that was created by a government, not a
people. A State created from the top down, not the

bottom up. The majority of India's citizens will not (to this day) be able to identify her boundaries on a map, or say which language is spoken where or which god is worshipped in what region. Most are too poor and too uneducated to have even an elementary idea of the extent and complexity of their own country. The impoverished, illiterate agrarian majority have no stake in the State. And indeed, why should they, how can they, when they don't even know what the State is? To them, India is, at best, a noisy slogan that comes around during the elections. Or a montage of people on Government TV programmes wearing regional costumes and saying *Mera Bharat Mahaan*.

The people who have a vital stake (or, more to the point, a business interest) in India having a single, lucid, cohesive national identity are the politicians who constitute our national political parties. The reason isn't far to seek, it's simply because their struggle, their career goal, is – and must necessarily be – to *become* that identity. To be identified with that identity. If there isn't one, they have to manufacture one and persuade people to vote for it. It isn't their fault. It comes with the territory. It is inherent in the nature of our system of centralized government. A congenital defect in our particular brand of democracy. The greater the numbers of illiterate people, the poorer the country and the more morally bankrupt the politicians, the cruder the ideas of what that identity should be. In a situation like this, illiteracy is not just sad, it's downright dangerous. However, to be fair, cobbling together a viable pre-digested 'National Identity' for India would be a formidable challenge even for the wise and the visionary. Every single Indian citizen could, if he or she wants to, claim to belong to some minority or the other. The fissures, if you look for them, run vertically, horizontally, and are layered, whorled, circular, spiral, inside out and outside in. Fires when they're lit race along any one of these schisms, and in the process, release tremendous bursts of political energy. Not unlike what happens when you split an atom.

It is this energy that Gandhi sought to harness when he rubbed the magic lamp and invited Ram and Rahim to partake of human politics and India's war of independence against the British. It was a sophisticated, magnificent, imaginative struggle, but its objective was simple and lucid, the target highly visible, easy to identify and succulent with political sin. In the circumstances, the energy found an easy focus. The trouble is that the circumstances are entirely changed now, but the genie is out of its lamp, and won't go back in. (It *could* be sent back, but nobody wants it to go, it's proved itself too useful.) Yes, it won us freedom. But it also won us the carnage of Partition. And now, in the hands of lesser statesmen, it has won us the Hindu Nuclear Bomb.

To be fair to Gandhi and to other leaders of the National Movement, they did not have the benefit of hindsight, and could not possibly have known what the eventual, long-term consequences of their strategy would be. They could not have predicted how quickly the situation would careen out of control. They could not have foreseen what would happen when they passed their flaming torches into the hands of their successors, or how venal those hands could be.

It was Indira Gandhi who started the real slide. It is she who made the genie a permanent State Guest. She injected the venom into our political veins. She invented our particularly vile local brand of political expediency. She showed us how to conjure enemies out of thin air, to fire at phantoms that she had carefully fashioned for that very purpose. It was she who discovered the benefits of never burying the dead, but preserving their putrid carcasses and trundling them out to worry old wounds when it suited her. Between herself and her sons she managed to bring the country to its knees. Our new Government has just kicked us over and arranged our heads on the chopping block.

The BJP is, in some senses, a spectre that Indira Gandhi and the Congress created. Or, if you want to be less harsh, a spectre that fed and reared itself in the political spaces and communal suspicion that the Congress nourished and cultivated. It has put a new complexion on the politics of governance. While Mrs Gandhi played hidden games with politicians and their parties, she reserved a shrill convent school rhetoric, replete with tired platitudes, to address the general public. The BJP, on the other hand, has chosen to light its fires directly on the streets and in the homes and hearts of people. It is prepared to do by day what the Congress would do only by night. To legitimize what was previously considered unacceptable (but done anyway). There is perhaps a fragile case to be made here in favour of hypocrisy. Could the hypocrisy of the Congress Party, the fact that it conducted its wretched affairs surreptitiously instead of openly, could that possibly mean there is a tiny glimmer of guilt somewhere? Some small fragment of remembered decency?

Actually, no.

No.

What am I doing? Why am I foraging for scraps of hope?

The way it has worked – in the case of the demolition of the Babri Masjid as well as in the making of the nuclear bomb – is that the Congress sowed the seeds, tended the crop, then the BJP stepped in and reaped the hideous harvest. They waltz together, locked in each other's arms. They're inseparable, despite their professed differences. Between them they have brought us here, to this dreadful, dreadful place.

The jeering, hooting young men who battered down the Babri Masjid are the same ones whose pictures appeared in the papers in the days that followed the nuclear tests. They were on the streets, celebrating India's nuclear bomb and simultaneously 'condemning Western Culture' by emptying crates of Coke and Pepsi into public drains. I'm a little baffled by their logic: Coke is Western Culture, but the nuclear bomb is an old Indian tradition?

Yes, I've heard – the bomb is in the Vedas. It might be, but if you look hard enough, you'll find Coke in the Vedas too. That's the great thing about all religious texts. You can find anything you want in them – as long as you know what you're looking for.

But returning to the subject of the non-Vedic 1990s: we storm the heart of whiteness, we embrace the most diabolical creation of Western science and call it our own. But we protest against their music, their food, their clothes, their cinema and their literature. That's not hypocrisy. That's humour.

It's funny enough to make a skull smile.

We're back on the old ship. The SS *Authenticity & Indianness*.

If there is going to be a pro-authenticity/anti-national drive, perhaps the Government ought to get its history straight and its facts right. If they're going to do it, they may as well do it properly.

First of all, the original inhabitants of this land were not Hindu. Ancient though it is, there were human beings on earth before there was Hinduism. India's Adivasi people have a greater claim to being indigenous to this land than anybody else, and how are they treated by the State and its minions? Oppressed, cheated, robbed of their lands, shunted around like surplus goods. Perhaps a good place to start would be to restore to them the dignity that was once theirs. Perhaps the Government could make a public undertaking that more dams like the Sardar Sarovar on the Narmada will not be built, that more people will not be displaced.

But, of course, that would be inconceivable, wouldn't it? Why? Because it's impractical. Because

Adivasis don't really matter. Their histories, their customs, their deities are dispensable. They must learn to sacrifice these things for the greater good of the Nation (that has snatched from them everything they ever had).

OK, so that's out.

For the rest, I could compile a practical list of things to ban and buildings to break. It'll need some research, but off the top of my head, here are a few suggestions.

They could begin by banning a number of ingredients from our cuisine: chillies (Mexico), tomatoes (Peru), potatoes (Bolivia), coffee (Morocco), tea, white sugar, cinnamon (China) . . . they could then move into recipes. Tea with milk and sugar, for instance (Britain).

Smoking will be out of the question. Tobacco came from North America.

Cricket, English and Democracy should be forbidden. Either kabaddi or kho-kho could replace cricket. I don't want to start a riot, so I hesitate to suggest a replacement for English (Italian ... ? It has found its way to us via a kinder route: marriage, not Imperialism). We have already discussed (earlier in this essay) the emerging, apparently acceptable alternative to democracy.

All hospitals in which Western medicine is practised or prescribed should be shut down. All national newspapers discontinued. The railways dismantled. Airports closed. And what about our newest toy – the mobile phone? Can we live without it, or shall I suggest that they make an exception there? They could put it down in the column marked 'Universal'? (Only essential commodities will be included here. No music, art or literature.)

Needless to say sending your children to university in the US, and rushing there yourself to have your prostate operated upon will be a cognizable offence.

The building demolition drive could begin with the Rashtrapati Bhavan and gradually spread from cities to the countryside, culminating in the destruction of all monuments (mosques, churches, temples) that were built on what was once Adivasi or forest land.

It will be a long, long list. It would take years of work. I couldn't use a computer because that wouldn't be very authentic of me, would it?

I don't mean to be facetious, merely to point out that this is surely the shortcut to hell. There's no such thing as an Authentic India or a Real Indian. There is no Divine Committee that has the right to sanction one single, authorized version of what India is or should be. There is no one religion or language or caste or region or person or story or book that can claim to be its sole representative. There are, and can only be, visions of India, various ways of seeing it – honest, dishonest, wonderful, absurd, modern, traditional, male, female. They can be argued over, criticized, praised, scorned, but not banned or broken. Not hunted down.

Railing against the past will not heal us. History has *happened*. It's over and done with. All we can do is to change its course by encouraging what we love instead of destroying what we don't. There is beauty yet in this brutal, damaged world of ours. Hidden, fierce, immense. Beauty that is uniquely ours and beauty that we have received with grace from others, enhanced, reinvented and made our own. We have to seek it out, nurture it, love it. Making bombs will only destroy us. It doesn't *matter* whether or not we use them. They will destroy us either way.

*

India's nuclear bomb is the final act of betrayal by a ruling class that has failed its people.

However many garlands we heap on our scientists, however many medals we pin to their chests, the truth is that it's far easier to make a bomb than to educate four hundred million people.

According to opinion polls, we're expected to believe that there's a national consensus on the issue. It's official now. Everybody loves the bomb. (Therefore the bomb is good.)

*

Is it possible for a man who cannot write his own name to understand even the basic, elementary facts about the nature of nuclear weapons? Has anybody told him that nuclear war has nothing at all to do with his received notions of war? Nothing to do with honour, nothing to do with pride. Has anybody bothered to explain to him about thermal blasts, radioactive fallout and the nuclear winter? Are there even words in his language to describe the concepts of enriched uranium, fissile material and critical mass? Or has his language itself become obsolete? Is he trapped in a time capsule, watching the world pass him by, unable to understand or communicate with it because his language never took into account the horrors that the human race would dream up? Does he not matter at all, this man? Shall we just treat him like some kind of a cretin? If he asks any questions, ply him with iodine pills and parables about how Lord Krishna lifted a hill or how the destruction of Lanka by Hanuman was unavoidable in order to preserve Sita's virtue and Ram's reputation? Use his own beautiful stories as weapons against him? Shall we release him from his capsule only during elections, and once he's voted, shake him by the hand, flatter him with some bullshit about the Wisdom of the Common Man, and send him right back in?

*

I'm not talking about one man of course, I'm talking about millions and millions of people who live in this country. This is their land too, you know. They have the right to make an informed decision about its fate and, as far as I can tell, nobody has informed them about anything. The tragedy is that nobody could, even if they wanted to. Truly, literally, there's no language to do it in. This is the real horror of India. The orbits of the powerful and the powerless spinning further and further apart from each other, never intersecting, sharing nothing. Not a language. Not even a country.

Who the hell conducted those opinion polls? Who the hell is the Prime Minister to decide whose finger will be on the nuclear button that could turn everything we love – our earth, our skies, our mountains, our plains, our rivers, our cities and villages – to ash in an instant? Who the hell is he to reassure us that there will be no accidents? How does he know? Why should we trust him? What has he ever done to make us trust him? What have any of them ever done to make us trust them?

*

The nuclear bomb is the most anti-democratic, anti-national, anti-human, outright evil thing that man has ever made.

If you are religious, then remember that this bomb is Man's challenge to God.

It's worded quite simply: *We have the power to destroy everything that You have created.*

If you're not (religious), then look at it this way. This world of ours is four thousand, six hundred million years old.

It could end in an afternoon.

August 1998